SpringerBriefs in Neuroscienc

For further volumes:
http://www.springer.com/series/8878

Caio Maximino

Serotonin and Anxiety

Neuroanatomical, Pharmacological, and Functional Aspects

Caio Maximino
Instituto de Ciencias Biologicas
Universidade Federal do Para
R. Augusto Correa 01
Belem 66085-110
Brazil

ISSN 2191-558X ISSN 2191-5598 (electronic)
ISBN 978-1-4614-4047-5 ISBN 978-1-4614-4048-2 (eBook)
DOI 10.1007/978-1-4614-4048-2
Springer New York Heidelberg Dordrecht London

Library of Congress Control Number: 2012939441

Printed on acid-free paper

Springer is part of Springer Science+Business Media (www.springer.com)

To Monica Gomes Lima, for putting up with my own anxieties during writing

Acknowledgment

First and foremost, this book is the result of the efforts of my father, Odair Gimenes Martins, in exposing me (sometimes in a brutally honest manner) to the pros and cons of an academic career. Without these efforts, I'd never have knowledge nor motivation to write work.

This work also resulted from extensive and fruitful discussions with my colleagues at the Neuroendocrinology Lab, as well as nights and days of conversations with my advisor, Anderson Manoel Herculano. For that, I'm indebted.

This book was written while under two scholarships from CAPES/Brazil, the first during my Master's and the second during the doctorate. Without this financial support, I'd never be able to have time to write this.

Contents

1 Introduction 1
1.1 Anxiety and Risk Assessment 2
1.2 Fear/panic and the Cerebral Aversive System 5
1.3 "Coping" Styles, Stress Reactivity, and the Active–Passive Continuum 6
References 8

2 Serotonin in the Nervous System of Vertebrates 15
2.1 Synthesis and Metabolism of Serotonin. 15
2.2 Transport of Serotonin: SERT and $Uptake_2$ 18
2.3 Serotonin Receptors 22
2.3.1 5-HT1A Receptors 23
2.3.2 $5\text{-}HT_{1B}$ Receptors 25
2.3.3 $5\text{-}HT_{2C}$ Receptors 26
References 27

3 Nodal Structures in Anxiety-Like and Panic-Like Responses 37
3.1 Nodal Structures Regulating Anxiety: The Behavioral Inhibition System 37
3.2 "Limbic" Portions of the Medial Prefrontal Cortex 37
3.3 The Extended Amygdala 40
3.4 The Ventral Hippocampus 42
3.5 The Lateral Habenula 46
3.6 Nodal Structures Regulating Panic: The Cerebral Aversive System 47
3.7 The Central Amygdala 48
3.8 The Medial Hypothalamic Defense System 51
3.9 The Mesopontine Rostromedial Tegmental Nucleus 53
3.10 The Periaqueductal Gray Area 53
3.11 Locus Coeruleus 57
References 60

4 The Deakin–Graeff Hypothesis 79
4.1 Destruction or Blockade of DRN Neurons is Anxiolytic and Panicogenic 80
4.2 The Defensive Context for Increased Serotonin Release 81
References 82

5 Topographic Organization of DRN 87
5.1 The Dorsal Portion of the DRN is Part of a Mesocorticolimbic System Involved in Anxiety-Like Responses 87
5.2 The Caudal Portion of the DRN is Highly Responsive to Stress-Related Peptides 91
5.3 The Lateral Wings of the DRN are Involved in Panic-Like Responses 93
References 96

6 General Conclusions 105
References 106

Index 109

Chapter 1
Introduction

Along with benzodiazepines, drugs targeting the serotonergic system represent the major class of anxiolytic drugs. Among available serotonergic drugs, selective serotonin reuptake inhibitors still represent the most prescribed treatment for anxiety disorders, even though they are associated with low efficacy in a considerable proportion of patients, a delayed onset of therapeutic action, and diverse collateral effects which reduce tolerance (e.g., sexual dysfunction, weight changes). There is considerable debate regarding the true contribution of serotonin or serotonin receptors to the therapeutic action of these drugs [1, 2], given that the acute increase in 5-HT concentrations in the synapse are not temporally correlated with the onset of therapeutic action.

The richness of the serotonergic system is reflected in the great quantity of receptor subtypes found in the brain [3]. This diversity underlines the possibility of different roles for each receptor subtype, and therefore to the potential for the production of more specific anxiolytic drugs. Nonetheless, efforts in the production of such drugs have resulted in disappointment [2]: with the exception of buspirone, a partial agonist at 5-HT_{1A} receptors that was introduced in the treatment of generalized anxiety disorder in 1985, no other anxiolytic agent targeting serotonin receptors produced robust clinical efficacy [4].

A distinction between anxiety and fear has been drawn on the basis of pharmacological dissociability (Table 1.1 [4–12]), neuroanatomical basis [8, 10, 13], and on its relation to stressor controllability [14, 15] and/or predictability [16–21]. These criteria are, of course, not mutually exclusive. In this book, we follow an integrated approach which considers defensive responses (A) as functions of discreteness of ambiguity of threat, defensive distance/predatory imminence continuum, and presence of particular environmental affordances [22, 23]; (B) controlled by different levels of a hierarchically organized behavioral inhibition system (anxiety-like responses) or cerebral aversive system (fear-like responses) [10]; and (C) differentially modulated by serotonergic neurotransmission [24] (Table 1.2).

C. Maximino, *Serotonin and Anxiety*, SpringerBriefs in Neuroscience,
DOI: 10.1007/978-1-4614-4048-2_1, © The Author(s) 2012

Table 1.1 Various classes of drugs vary in clinical efficacy in the treatment of anxiety disorders

Disorder	BZD	Triazolo	Bus	Imi	Clom	MAOi	SSRI	SARI	β
GAD	↓	↓	↓	↓	↓	0	↓	↓	0
Panic	0	↓	0	↓	↓↓	↓	↓	?	0
PTSD	0	?	0	↓	?	↓	↓	?	?
Simple phobia	0	?	?	0	?	(↓)	(↓)	?	0
Social anxiety	↓	(↓)	(↓)	0	(↓)	↓	↓	↓	↓
OCD	0	0	(↓)	(↓)	↓↓	(↓)	↓↓	↓	0

BZD benzodiazepine, *Triazolo* triazolo-benzodiazepines, *Bus* buspirone, *Imi* imipramine, *Clom* clomipramine, *MAOi* monoamine oxidase inhibitor, *SSRI* selective serotonin reuptake inhibitor, *SARI* serotonin antagonist and reuptake inhibitor, *β* β-adrenoceptor antagonist
Symbols ↓ symptom decrease, ↓↓ major symptom decrease, (↓) contradictory or insufficient findings, 0: no clinical efficacy, ?: clinical efficacy not assayed
Adapted from Refs. [4–12]

1.1 Anxiety and Risk Assessment

Anxiety is a state of "action readiness" associated with unpredictable or uncontrollable aversive stimuli [14, 16–21, 25]. "Readiness" here implies preparedness for action *if* and *when* appropriate conditions (affordances) arise [26, 27]. In a situation of uncertain or merely probable risk (called "pre-encounter environment" by Fanselow and colleagues [23, 28]), behavioral adjustments grouped under the general category of "risk assessment" are made. Risk assessment is a collection of adjustments that is involved in detection and analysis of threat stimuli and the context in which it occurs [29]. Thus, animals will shift attention from ongoing motivated behavior toward detecting and/or responding to potential predators. In situations of uncertainty regarding risk, animals adopt a baseline of "apprehension", leading to the selection of vigilant behaviors [22, 29–33]. In such situations, animals also tend to "overestimate" the actual level of threat; this "cognitive bias" [34] leads animals to inhibit ongoing behavior and flee, hide or freeze if any signal of risk is detected. Depending on environmental affordances, animals tend to retreat to protected areas [35], resort to thigmotaxis ("wall-hugging") [36] and scototaxis ("dark preference") [37], and establish "home bases" to which they constantly return after exploring the environment [38].

An important environmental configuration which leads to risk assessment behavior is novelty. Montgomery [39] proposed that novel environments evoke both exploratory drives and fear, producing an approach-avoidance conflict. Importantly, novelty is a situation of potential risk, and exploratory behavior is adjusted accordingly. This is explored in diverse behavioral models of anxiety, in which the forced exposure to novelty leads to risk assessment behavior and adjustments of exploration (thigmotaxis, scototaxis, refuge use, home base behavior). In totally novel environments, anxiolytic drugs increase exploratory behavior, particularly of aversive portions of the apparatuses (e.g., open arms of an elevated plus-maze, lit chamber of a light/dark box, center of an open-field) [40].

Table 1.2 Stimulus control of defensive behavior, in relation to threat source, associated level in the predatory imminence continuum, and environmental affordances

Source of threat	Predatory imminence	Affordance	Behavior	Neuroanatomy
Uncertain	Pre-encounter		Risk assessment	Medial prefrontal cortex Septo-hippocampal system Extended amygdala Lateral habenula
		Walls, refuges	Adjustment of exploratory behavior	Cingulate cortex Septo-hippocampal system Extended amygdala Lateral habenula
			Cognitive bias	Septo-hippocampal system Extended amygdala Lateral habenula
			Behavioral activation	Mesolimbic dopaminergic system
			Attention/arousal	Cortico-coerulear projection
Discrete	Post-encounter	Escape route available	Flight	Medial hypothalamic defense system Dorsal PAG RMTg
		No escape route	Freezing	PAG RMTg
		Conspecifics nearby	Alarm call/USV	Dorsal PAG
		Hiding places available	Hiding	
			Neurovegetative adjustments	LH PVN
	Circa-strike		Defensive fight	
			Analgesia	PAG
	Predator contact		Startle	Elementary startle circuit
			Tonic immobility	Ventral PAG

Also marked are the brain regions most likely to be involved in the control of such behavior
LH lateral hypothalamus, *PAG* periaqueductal gray area, *PVN* paraventricular nucleus, *RMTg* rostromedial tegmental area, *USV* ultrasonic vocalization
Adapted from Refs. [8, 10, 13–23, 25, 28]

In situations where animals are allowed to freely *choose* between a novel and a familiar environment, nonetheless, they tend to prefer novelty [41], and this preference is reversed by anxiolytic drugs [42–44]. Likewise, if an animal is re-exposed to the elevated plus-maze 24 h after the first exposure ("trial 2"), time spent in the open arms is further decreased, and response to anxiolytic drugs

(5-HT_{1A}R agonists, benzodiazepines) is eliminated; this "one-trial tolerance" effect has been proposed as a model of simple phobia [45]. Interestingly, administration of D-cycloserine, a partial agonist at the glycine_B site of the NMDA receptor, at the end of trial 1 potentiates the increase in open arm avoidance, without reverting the effect on benzodiazepine efficacy [46].

Anxiety-like behavior can also be observed in the home cage after the administration of anxiogenic drugs, including benzodiazepine inverse agonists and antagonists, caffeine, yohimbine, corticotropin releasing factor (CRF), and *m*-chlorophenyl piperazine (mCPP). After administration of such drugs, animals engage in spontaneous non-ambulatory motor activity (SNAMA, part of the class of risk assessment behavior), including visual scanning of the environment, head movements associated with sniffing, and shifts in body position, for up to 90 min. [47, 48].

Anxiety-like behavior either at novel environments or at the home cage can be increased by stressful manipulations [49]. This "fear potentiation" reflects an enhanced anxiety *state* in face of an allostatic situation, and can last from 90 min to 3 weeks, depending on the stressor used (immobilization, electrical shocks, exposure to predators or partial predator stimuli, social defeat, etc.). In these cases, enhanced secretion of corticosteroids by the adrenal glands facilitates the expression of CRF in the central amygdala and bed nucleus of the stria terminalis, leading to increased anxiety, increased norepinephrine release in the locus coeruleus, and increases in the extracellular concentrations of serotonin in limbic regions. It has been suggested that, while short-term effects of mild stressors can model acute allostatic situations, the long lasting effects of predator exposure on defensive behavior is a good model of post-traumatic stress disorder [50].

A manipulation which induces long-term increases in anxiety-like behavior is acute uncontrollable stress, producing effects that last up to 24–72 h [14, 16, 17, 51–55]. When animals are exposed to electric shocks which are contingent to escape responses, they quickly develop "active coping" behavior; if, however, electric shocks are not contingent to escape (i.e., they are inescapable or uncontrollable), these animals develop "passive coping" behavior, freezing rather than attempting to escape [54, 56]; show higher corticosteroid release than animals which have been exposed to escapable shock [57]; and show facilitated conditioning of fear and impairment of escape [58] and increased anxiety in an elevated plus-maze [57]. This sensitized state has been termed "learned helplessness" by earlier theorists, and the lack of control over the aversive event has been proposed as an important component of anxiety disorders [59–61]. Similar effects are observed in animals which have been exposed to chronic unpredictable stress (CUS [62, 63], but the effects of this latter manipulation on anxiety-like behavior are controversial (e.g., effects on the EPM or LD test are not always observed [64, 65]) and are not immediate, with a delay of about a week for onset [65].

1.2 Fear/panic and the Cerebral Aversive System

Although some authors (e.g., Panksepp [66, 67]) have argued that anxiety is a "minor fear", being controlled by essentially the same neural structures which control fear- and panic-like responses, there is now considerable evidence that these responses are controlled by different (albeit interrelated) systems [8, 22]. Fear is a response to discrete threatening stimuli, and controlled by diencephalic, mesencephalic, and rhombencephalic structures which comprise a "fight/flight/freeze" or "cerebral aversive" system" [10]. The fight/flight response represents the active/proactive behavioral patterns that were first characterized by Cannon [68], and is usually accompanied by neurovegetative adjustments that are mediated by the autonomic nervous system and the hypothalamic-pituitary-adrenal axis. Conversely, freezing, characterized by the absence of all movement except breathing, is a strategy to avoid detection [28, 69, 70] and therefore conditional on the source of danger being present but distant and on the unavailability of escape routes or hiding places (Table 1.2). In addition to these responses, analgesia is also a consequence of fear, allowing the animal to express defensive behaviors and waive recuperative behaviors in the presence of danger [71].

At closer levels of predatory imminence, fight-or-flight mechanisms are urgently recruited, leading to a rapid burst of activity. If the animal is cornered, defensive threat ensues; cornered rats, for example, vocalize, bare its teeth, and attempt to jump beyond or at the predator [35], and marmosets exposed to a predator vocalize and sway [72]. If escape is not available and defensive threat is ineffective, animals resort to defensive fighting [73]; likewise, when two animals are placed in a cage where they receive inescapable foot shocks, fighting emerges after a number of presentations [74].

Startle responses have been reported in the literature, and can be enhanced/sensitized by conditioned fear states [75]. Blanchards [22] proposed that the function of these startle response is to "scare" a predator; a more parsimonious explanation is that aversive events sensitize brain systems which increase the organism's propensity for flight, thus leading to fear-potentiated startle [76]. The neuroanatomical underpinnings of fear-potentiated startle responses have been extensively studied by Michael Davis and collaborators [13, 77–82].

In the present time, one of the most widely used paradigms in the study of fear responses is Pavlovian fear conditioning [83]—or, more precisely, "conditional fear". "Cue" fear conditioning leads to the generation of defensive reactions to stimuli which predict aversive consequences, and can be a simple conditioned stimulus (CS)-unconditioned stimulus (US) pairing, or more complex arrangements, such as discriminative fear conditioning (in which a given CS [CS^D] is paired with an aversive US, while another CS [CS^Δ] is not) or ambiguous cue fear conditioning (in which a given CS is not reliably paired with the aversive US). Contextual fear conditioning, on the other hand, refers to the degree with which fear responses can be elicited by the environment in which conditioning took place; in this perspective, contextual fear conditioning has been proposed as a model of anxiety [84]. There are many aspects of contextual fear conditioning, however, that are contrary to this interpretation [83]. The most important of these features is that fear conditioning (even contextual fear conditioning) is under immediate control of the level of threat in the environment [85–89] so that, for example, the degree of freezing to context is proportional to shock intensity and number of trials [90, 91]. The principal argument underlying the proposal that contextual fear conditioning is a model of anxiety is that freezing to context is a response to "diffuse" signals of threat, as opposed to the discrete signals which are thought to mediate anxiety [92]; moreover, contextual fear conditioning hinders the evocation of fear responses from the dorsal periaqueductal gray area [93], and freezing to context is reduced by benzodiazepines, 5-HT_{1A} receptor agonists, selective serotonin reuptake inhibitors, and monoamine oxidase inhibitors [94–96]; while the first three groups of drugs are effective in the clinical management of generalized anxiety disorder, MAO inhibitors are not (Table 1.1), somewhat weakening the argument.

1.3 "Coping" Styles, Stress Reactivity, and the Active–Passive Continuum

The notion that behavioral responses to threatening stimuli can be "active" (escape, defensive fighting) or "passive" (freezing, risk assessment), depending on the level of predator imminence is central on the modern view of fear/anxiety. It has been argued that an active–passive continuum underlies individual differences in the behavioral and neuroendocrinological responses to stress in animals [97, 98]. In this view, behavior naturally varies in at least two independent dimensions, the quality of responses to allostatic conditions (coping styles) and the quantity of such responses (stress reactivity) [99]. The concept of coping style, in the context of defensive behavior, was derived from the observation that individual variations in offensive aggressive behavior in rodents is stable over time and correlated with responses in a variety of allostatic conditions [97]. Thus, animals selected for short attack latencies in a resident-intruder paradigm, or animals selected for high learning of active avoidance, show increased time spent burying a probe in the defensive burying paradigm, while animals selected for long attack latencies or low learning of active avoidance show increased freezing duration in the same test [100, 101]. Likewise, aggressive animals (i.e., animals with short attack latencies) react with active swimming and climbing in the forced swim test, while non-aggressive animals show predominantly floating behavior [102]. However, these animals show no differences in the time spent on the open arm of an elevated plus-maze [103]; Koolhaas and colleagues [97] argued that a correlation with aggressive behavior can be observed only in behavioral tasks which allow animals to choose between active and passive behaviors.

From a functional neuroanatomical point of view, it is not yet clear whether these coping patterns are controlled by distinct subpopulations of neurons, or by the same population modulated by different degrees of neuromodulation. The important question here is where the "switch" lies. On the grounds of lesion and stimulation experiments, it has been argued that the amygdala is associated with the active coping, while the septo-hippocampal system is associated with passive coping [104]. This view is unlikely. It has been demonstrated that, in the resident-intruder paradigm, aggressive animals show an increased number of c-Fos-positive neurons than non-aggressive animals in the central amygdala, anterior bed nucleus of the stria terminalis, ventrolateral hypothalamus, nucleus accumbens shell, orbital frontal cortex, ventrolateral periaqueductal gray, while in the lateral septum and dorsolateral periaqueductal gray non-aggressive animals show more c-Fos-like immunoreactivity than aggressive animals; however, both groups showed increased c-Fos-like immunoreactivity in these regions than animals which have not been exposed to intruding conspecifics [105–107]. Likewise, when mice are exposed to an innately aversive ultrasound ($\sim$22 kHz), they tend to flight; if they are previously exposed to footshocks 24 h before ultrasound exposure, this behavior shifts to freezing [108]. In mice which have not been sensitized with footshocks (i.e., animals predominantly displaying flight), c-Fos-like immunoreactivity is observed in the medial prefrontal cortex, medial amygdala, shell of the nucleus accumbens, diagonal band of Broca, anterior portion of the bed nucleus of the stria terminalis, lateral and posterior hypothalamus, and in the mesopontine rostromedial tegmental nucleus. On the other hand, on mice which have been sensitized (i.e., animals predominantly displaying freezing), c-Fos-like immunoreactivity is observed in the lateral septum and in the medial and paraventricular hypothalamic nuclei. Again, all these regions were activated on both groups in relation to control animals [108]. Conspicuously, lacking in the stereological analysis of Mongeau and colleagues [108] is a differentiation between subnuclei. Thus, while the central nucleus of the amygdala has been shown to be differentially activated in both groups in relation to control animals, no difference is reported between sensitized and non-sensitized animals. However, from their Fig. 3.5a, it can be inferred that non-sensitized animals show more c-Fos-like immunoreactivity in the medial portion of the central amygdala, while sensitized mice show more c-Fos-like immunoreactivity in the lateral and capsular portions of the central amygdala. This distinction is important, as serotonergic mechanisms in the subregions of the central amygdala have been implicated in "switching" from passive to active coping styles [109].

An important role for the serotonergic "tone" has been proposed in the control of coping styles. S-15535, a full agonist at 5-HT$_{1A}$ autoreceptors and antagonist at postsynaptic 5-HT$_{1A}$ binding sites, reduces aggressive behavior in rats, an effect which is 15 times higher in high-aggressive than low-aggressive animals [110]. Likewise, in the forced swim test 5-HT$_{1A}$ agonists reduce immobility and increase escape attempts in passive animals, and having opposite effect on active animals [102]. This difference in effect is indicative of increased inhibitory tone in high-aggressive animals, and De Boear and Koolhaas suggested that it can explain the negative correlation between baseline 5-HT levels and aggression that is found in many species.

It has also been proposed that these behavioral strategies can be mediated by the bimodal effects of corticotropin-releasing factor (CRF) on serotonin release by the dorsal raphe nucleus [111, 112]. Acting on CRF_1 receptors, CRF inhibits serotonin release, while the activation of CRF_2 receptors increase it [113–120]; CRF itself has more affinity for CRF_1 than CRF_2 receptors [115], which led Valentino and colleagues [111, 121, 122] to propose that the activation of the first leads to active coping strategies, and sustained stress would lead to activation of the latter and desensitization of the first and, consequently, to passive strategies.

These results suggest that both "trait" (i.e., interindividual differences) and "state" (i.e., stress-induced intraindividual differences) variation in coping styles are regulated by a set of structures that "switch" between active and passive behavioral responses. Importantly, serotonin release in these structures seems to be important in defining these strategies. It is not yet clear how these coping styles relate to anxiety- and fear-like behavior, but they certainly relate to responses to stress which can reflect on defensive responses [97, 98]. Thus, the role of serotonin in "switching" response modes will be a recurrent theme in this book.

References

1. Davis LL, Yonkers KA, Trivedi M, Kramer GL, Petty F (1999) The mechanism of action of SSRIs: a new hypothesis. In: Stanford SC (ed) Selective serotonin reuptake inhibitors (SSRIs): past, present and future. R. G. Landes Company, Austin, pp 171–185
2. Guimarães FS, Carobrez AP, Graeff FG (2008) Modulation of anxiety behaviors by 5-HT-interacting drugs. In: Blanchard RJ, Blanchard DC, Griebel G, Nutt D (eds) Handbook of anxiety and fear. Elsevier B. V., Amsterdam, pp 241–268
3. Barnes NM, Sharp T (1999) A review of central 5-HT receptors and their function. Neuropharmacology 38:1083–1152
4. Rickels K, Rynn M (2002) Pharmacotherapy of generalized anxiety disorder. J Clin Psychiatry 63:9–16
5. McNaughton N, Panickar KS, Logan B (1996) The pituitary-adrenal axis and the different behavioral effects of buspirone and chlordiazepoxide. Pharmacol Biochem Behav 54:51–56
6. Baldwin DS, Bridle D, Ekelung A (2003) Pharmacotherapy of anxiety disorders. In: Kasper S, den Boer JA, Ad Sitsen JM (eds) Handbook of depression and anxiety, 2nd edn. Revised and expanded. Marcel Dekker, Inc., New York, pp 732–756
7. Blanchard RJ, Yudko EB, Rodgers RJ, Blanchard DC (1993) Defense system psychopharmacology: an ethological approach to fear and anxiety. Behav Brain Res 58:155–165
8. Gray JA, McNaughton N (2000) Neuropsychology of anxiety: an enquiry into the functions of the septo-hippocampal system, 2a edn. Oxford University Press, Oxford
9. Holmes A (2008) The pharmacology of anxiolysis. In: Blanchard RJ, Blanchard DC, Griebel G, Nutt D (eds) Handbook of anxiety and fear. Elsevier B. V., Amsterdam, pp 355–361
10. McNaughton N, Corr PJ (2004) A two-dimensional neuropsychology of defense: fear/anxiety and defensive distance. Neurosci Biobehav Rev 28:285–305

11. Stein DJ, Hollander E, Mullen LS, DeCaria CM, Liebowitz MR (1992) Comparison of clomipramine, alprazolam and placebo in the treatment of obsessive-compulsive disorder. Hum Psychopharmacol 7:389–395
12. Westenberg HGM (1999) Facing the challenge of social anxiety disorder. Eur Neuropsychopharmacol 9:S93–S99
13. Davis M, Shi C (1999) The extended amygdala: are the central nucleus of the amygdala and the bed nucleus of the stria terminalis differentially involved in fear versus anxiety? Ann N Y Acad Sci 877:281–291
14. Maier SF, Lr Watkins (1998) Stressor controllability, anxiety, and serotonin. Cogn Ther Res 22:595–613
15. Mowrer OH, Viek P (1948) An experimental analogue of fear from a sense of helplessness. J Abnorm Soc Psychol 83:193–200
16. Mineka S, Kihlstrom JF (1978) Unpredictable and uncontrollable events: a new perspective on experimental neurosis. J Abnorm Psychol 87:256–271
17. Seligman MEP, Maier SF, Solomon RL (1971) Unpredictable and uncontrollable aversive events. In: Brush FR (ed) Aversive conditioning and learning. Academic, New York
18. Clinchy M, Schulkin J, Zanette LY, Sheriff MJ, McGowan PO, Boonstra R (2011) The neurological ecology of fear: insights neuroscientists and ecologists have to offer one another. Front Behav Neurosci 5:Article 21
19. Rodgers RJ (1997) Animal models of 'anxiety': where next? Behav Pharmacol 8:477–496
20. Steimer T (2002) The biology of fear- and anxiety-related behaviors. Dialogues Clin Neurosci 4:231–249
21. Grillon C (2008) Models and mechanisms of anxiety: evidence from startle studies. Psychopharmacology 199:421–437
22. Blanchard DC, Blanchard RJ (2008) Defensive behaviors, fear, and anxiety. In: Blanchard RJ, Blanchard DC, Griebel G, Nutt D (eds) Handbook of fear and anxiety. Elsevier B. V., Amsterdam, pp 63–79
23. Fanselow M, Lester L (1988) A functional behavioristic approach to aversively motivated behavior: predatory imminence as a determinant of the topography of defensive behavior. In: Bolles RC, Beecher MD (eds) Evolution and learning. Erlbaum, Hillsdalse, pp 185–211
24. Deakin JFW, Graeff FG (1991) 5-HT and mechanisms of defence. J Psychopharmacol 5:305–315
25. Lowry CA, Hale MW (2010) Serotonin and the neurobiology of anxious states. In: Muller CP, Jacobs BL (eds) The behavioral neurobiology of serotonin. Academic, London, pp 379–397
26. Fridja NH (1986) The emotions. Cambride University Press, Cambridge
27. Fridja NH (2007) The laws of emotion. Erlbaum, Mahwah
28. Godsil BP, Tinsley MR, Fanselow MS (2003) Motivation. In: Healy AF, Proctor RW (eds) Experimental psychology. Handbook of psychology. Wiley, Hoboken, pp 33–60
29. Blanchard DC, Griebel G, Pobbe R, Blanchard RJ (2011) Risk assessment as an evolved threat detection and analysis process. Neurosci Biobehav Rev 35:991–998
30. Maximino C, Brito TM, Gouveia A Jr (2010) Construct validity of behavioral models of anxiety: where experimental psychopathology meets ecology and evolution. Psychol Neurosci 3:117–123
31. Brown JS, Kotler BP, Bouskila A (2001) Ecology of fear: foraging games between predator and prey with pulsed resources. Ann Zool Fenn 38:71–87
32. Kavaliers M, Choleris E (2001) Antipredator responses and defensive behavior: ecological and ethological approaches for the neurosciences. Neurosci Biobehav Rev 25:577–586
33. Brown JS, Landre JW, Gurung M (1999) The ecology of fear: optimal foraging, game theory and trophic interactions. J Mammal 80:385–399
34. Mendl M, Burman OHP, Parker RMA, Paul ES (2009) Cognitive bias as an indicator of animal emotion and welfare: emerging evidence and underlying mechanisms. Appl Anim Behav Sci 118:161–181

35. Blanchard RJ, Blanchard DC (1989) Antipredator defensive behaviors in a visible burrow system. J Comp Psychol 103:70–82
36. Creed RP Jr, Miller JR (1990) Interpreting animal wall-following behaviour. Experientia 46:758–761
37. Bourin M, Hascöet M (2003) The mouse light/dark box test. Eur J Pharmacol 463:55–65
38. Eilam D, Golani I (1989) Home base behavior of rats (*Rattus norvegicus*) exploring a novel environment. Behav Brain Res 34:199–211
39. Montgomery KC (1955) The relation between fear induced by novel stimulation and exploratory behavior. J Comp Physiol Psychol 48:254–260
40. Toth M, Zupan B (2007) Neurobiology of anxiety. In: Sibley DR, Hanin I, Kuhar M, Skolnick P (eds) Handbook of contemporary neuropharmacology, vol 2. Wiley, New York, pp 3–58
41. Hughes RN (1968) Behaviour of male and female rats with free choice of two environments differing in novelty. Anim Behav 13:30–32
42. Hughes RN (1972) Chlordiazepoxide modified exploration in rats. Psychopharmacologia 24:462–469
43. Hughes RN (1981) Oxprenolol and diazepam effects on activity, novelty preference and timidity in rats. Life Sci 29:1089–1092
44. Hughes RN, Greig AM (1975) Chlordiazepoxide effects on reactions to novelty and activity with and without prior drug experience. Psychopharmacology 42:289–292
45. File SE, Zangrossi H Jr, Viana M, Graeff FG (1993) Trial 2 in the elevated plus-maze: a different form of fear? Psychopharmacology 111:491–494
46. Rodgers RJ, Harvest H, Hassall C, Kaddour LA (2011) D-cycloserine enhances memory consolidation in the plus-maze retest paradigm. Behav Neurosci 125:106–116
47. Abrams JK, Johnson PL, Hay-Schmidt A, Mikkelsen J, Shekhar A, Lowry CA (2005) Serotonergic systems associated with arousal and vigilance behaviors following administration of anxiogenic drugs. Neuroscience 133:983–997
48. Lowry CA, Hale MW, Plant A, Windle RJ, Shanks N, Wood SA, Ingram CD, Renner KJ, Lightman SL, Summers CH (2009) Fluoxetine inhibits corticotropin-releasing factor (CRF)-induced behavioural responses in rats. Stress 12:225–239
49. Korte SM, De Boer SF (2003) A robust animal model of state anxiety: fear-potentiated behaviour in the elevated plus-maze. Eur J Pharmacol 463:163–175
50. Adamec R, Holmes A, Blundell J (2008) Vulnerability to lasting anxiogenic effects of brief exposure to predator stimuli: sex, serotonin and other factors—relevance to PTSD. Neurosci Biobehav Rev 32:1287–1292
51. Edwards E, Johnson J, Anderson D, Turano P, Henn FA (1986) Neurochemical and behavioral consequences of mild, uncontrollable shock: effects of PCPA. Pharmacol Biochem Behav 25:415–421
52. Maier SF, Watkins LR (2005) Stressor controllability and learned helplessness: the roles of the dorsal raphe nucleus, serotonin, and corticotropin-releasing factor. Neurosci Biobehav Rev 29:829–841
53. Zvolensky MJ, Lejuez CW, Eifert GH (2000) Prediction and control: operational definitions for the experimental analysis of anxiety. Behav Res Ther 38:653–663
54. Seligman ME, Maier SF (1967) Failure to escape traumatic shock. J Exp Psychol 74:1–9
55. Weiss JM (1968) Effects of coping responses on stress. J Comp Physiol Psychol 65:251–260
56. Overmier JB, Seligman ME (1967) Effects of inescapable shock upon subsequent escape and avoidance responding. J Comp Physiol Psychol 63:28–33
57. Korte SM, Bouws GA, Bohus B (1992) Adrenal hormones in rats before and after stress-experience: effects of ipsapirone. Physiol Behav 51:1129–1133
58. Maier SF (1990) Role of fear in mediating shuttle escape learning deficit produced by inescapable shock. J Exp Psychol: Anim Behav Process 16:137–149
59. Barlow DH (2002) Anxiety and its disorders: the nature and treatment of anxiety and panic, 2nd edn. Guilford Press, New York

60. Seligman MEP (1975) Helplessness: on depression, development, and death. Freeman, San Francisco
61. Maier SF (1993) Learned helplessness, fear and anxiety. In: Stanford P, Solomon K (eds) Stress: from synapse to syndrome. Academic, London, pp 207–248
62. Bondi CO, Rodriguez G, Gould GG, Frazer A, Morilak DA (2008) Chronic unpredictable stress induces a cognitive deficit and anxiety-like behavior in rats that is prevented by chronic antidepressant drug treatment. Neuropsychopharmacology 33:320–331
63. Piato ÑL, Capiotti KM, Tamborski AR, Oses JP, Barcellos LJG, Bogo MR, Lara DR, Vianna MR, Bonan CD (2011) Unpredictable chronic stress model in zebrafish (*Danio rerio*): behavioral and physiological responses. Prog Neuropsychopharmacol Biol Psychiatry 35:5610567
64. Vyas A, Chattarji S (2004) Modulation of different states of anxiety-like behavior by chronic stress. Behav Neurosci 118:1450–1454
65. Matuszewich L, Karney JJ, Carter SR, Janasik SP, O'Brien JL, Friedman RD (2007) The delayed effects of chronic unpredictable stress on anxiety measures. Physiol Behav 90: 674–681
66. Panksepp J (1998) Affective neuroscience: the foundations of human and animal emotions. Oxford University Press, New York
67. Panksepp J (2004) Emerging neuroscience of fear and anxiety: therapeutic practice and clinical implications. In: Panksepp J (ed) Textbook of biological psychiatry. Wiley, New York, pp 489–519
68. Cannon WB (1915) Bodily changes in pain, hunger, fear and rage. Appleton, New York
69. Bolles RC, Riley AL (1972) Freezing as an avoidance response: another look at the operant-respondent distinction. Learn Motiv 4:268–275
70. Eilam D (2005) Die hard: a blend of freezing and fleeing as a dynamic defense—implications for the control of defensive behavior. Neurosci Biobehav Rev 29:1181–1191
71. Bolles RC, Fanselow M (1980) A perceptual-defensive-recuperative model of fear and pain. Behav Brain Sci 3:291–323
72. Barros M, Silva MAD, Huston JP, Tomaz C (2004) Multibehavioral analysis of fear and anxiety before, during, and after experimentally induced predatory stress in *Callithrix* penicillata. Pharmacol Biochem Behav 78:357–367
73. Blanchard DC, Blanchard RJ (1988) Ethoexperimental approaches to the biology of emotion. Annu Rev Psychol 39:43–68
74. Fanselow MS, Sigmundi RA (1982) The enhancement and reduction of defensive fighting by naloxone pretreatment. Physiol Psychol 10:313–316
75. Brown JS, Kalish HI, Farber IE (1961) Conditioned fear as revealed by magnitude of startle response to an auditory stimulus. J Exp Psychol 41:317–328
76. Koch M, Schnitzler H-U (1997) The acoustic startle response in rats—circuits mediating evocation, inhibition and potentiation. Behav Brain Res 89:35–49
77. Campeau S, Davis M (1995) Involvement of the central nucleus and basolateral complex of the amygdala in fear conditioning measured with fear-potentiated startle in rats trained concurrently with auditory and visual conditioned stimuli. J Neurosci 15:2301–2311
78. Miserendino MJ, Sananes CB, Melia KR, Davis M (1990) Blocking of acquisition but not expression of conditioned fear-potentiated startle by NMDA antagonists in the amygdala. Nature 345:716–718
79. Walker DL, Davis M (1997) Double dissociation between the involvement of the bed nucleus of the stria terminalis and the central nucleus of the amygdala in light-enhanced versus fear-potentiated startle. J Neurosci 17:9375–9383
80. Walker DL, Toufexis DJ, Davis M (2003) Role of the bed nucleus of the stria terminalis versus the amygdala in fear, stress, and anxiety. Eur J Pharmacol 463:199–216
81. Gewirtz JC, McNIsh KA, Davis M (1998) Lesions of the bed nucleus of the stria terminalis block sensitization of the acoustic startle reflex produced by repeated stress, but not fear-potentiated startle. Prog Neuropsychopharmacol Biol Psychiatry 22:625–648

82. Hitchcock JM, Davis M (1986) Lesions of the amygdala, but not of the cerebellum or red nucleus, block conditioned fear as measured with the potentiated startle paradigm. Behav Neurosci 100:11–22
83. Fanselow MS, Ponnusamy R (2008) The use of conditioning tasks to model fear and anxiety. In: Blanchard RJ, Blanchard DC, Griebel G, Nutt D (eds) Handbook of anxiety and fear. Elsevier B. V., Amsterdam, pp 29–48
84. Landeira-Fernandez J (1996) Context and pavlovian fear conditioning. Braz J Medial Biol Res 29:149–173
85. Bolles RC, Collier AC (1976) The effect of predictive cues on freezing in rats. Anim Learn Behav 4:6–8
86. Fanselow MS (1980) Conditioned and unconditioned components of post-shock freezing. Pavlovian J Biol Sci 15:177–182
87. Blanchard RJ, Fukunaga KK, Blanchard DC (1976) Environmental control of defensive reactions to footshock. Bull Psychon Soc 8:129–130
88. Rescorla RA (1968) Probability of shock in the presence and absence of CS in fear conditioning. J Comp Physiol Psychol 66:1–5
89. Young SL, Fanselow MS (1992) Associative regulation of pavlovian fear conditioning: unconditional stimulus intensity, incentive shifts, and latent inhibition. J Exp Psychol: Anim Behav Process 18:400–413
90. Fanselow MS, Tighe TJ (1988) Contextual conditioning with massed versus distributed unconditional stimuli in the absence of explicit conditional stimuli. J Exp Psychol: Anim Behav Process 14:187–199
91. Fanselow MS, Bolles RC (1979) Naloxone and shock-elicited freezing in the rat. J Comp Physiol Psychol 93:736–744
92. Lang PJ, Davis M, Öhman A (2000) Fear and anxiety: animal models and human cognitive psychophysiology. J Affect Disord 61:137–159
93. Magierek V, Ramos PL, Silveira-Filho NG, Nogueira RL, Landeira-Fernandez J (2003) Context fear conditioning inhibits panic-like behavior elicited by electrical stimulation of dorsal periaqueductal gray. NeuroReport 14:1641–1644
94. Guscott MR, Cook GP, Bristow LJ (2000) Contextual fear conditioning and baseline startle responses in the rat fear-potentiated startle test: a comparison of benzodiazepine/ γ-aminobutyric acid-a receptor agonists. Behav Pharmacol 11:495–504
95. Hashimoto S, Inoue T, Koyama T (1996) Serotonin reuptake inhibitors reduce conditioned fear stress-induced freezing behavior in rats. Psychopharmacology 123:182–186
96. Maki Y, Inoue T, Izumi T, Muraki I, Ito K, Kitaichi Y, Li X, Koyama T (2000) Monoamine oxidase inhibitors reduce conditioned fear stress-induced freezing behavior in rats. Eur J Pharmacol 406:411–418
97. Koolhaas JM, De Boer SF, Coppens CM, Buwalda B (2010) Neuroendocrinology of coping styles: towards understanding the biology of individual variation. Front Neuroendocrinol 31:307–321
98. Koolhaas JM, Korte SM, De Boer SF, van der Vegt BJ, van Reenen CG, Hopster H, de Jong IC, Ruis MAW, Blokhuis HJ (1999) Coping styles in animals: current status in behavior and stress-physiology. Neurosci Biobehav Rev 23:925–935
99. Steimer T, La Fleur FS, Schulz PE (1997) Neuroendocrine correlates of emotional reactivity and coping in male rats from the Roman high (RHA/Verh)- and low (RLA/Verh)-avoidance lines. Behav Genet 27:503–512
100. Sluyter F, Korte SM, Bohus B, van Oortmerssen GA (1996) Behavioral stress response of genetically selected aggressive and nonaggressive wild house mice in the shock-probe/ defensive burying test. Pharmacol Biochem Behav 54:113–116
101. Boersma GJ, Scheurink AJ, Wielinga PY, Steimer T, Benthem L (2009) The passive coping Roman low avoidance rat, a non-obese rat model for insulin resistance. Physiol Behav 97:353–358
102. Veenema AH, Cremers TI, Jongsma ME, Steenbergen PJ, De Boer SF, Koolhaas JM (2005) Differences in the effects of 5-HT(1A) receptor agonists on forced swimming behavior and

brain 5-HT metabolism between low and high aggressive mice. Psychopharmacology 178:151–160
103. De Boer SF, Koolhaas JM (2003) Defensive burying in rodents: ethology, neurobiology and psychopharmacology. Eur J Pharmacol 463:145–161
104. Henry JP, Stephens PM (1977) Stress, health and the social environment: a sociobiological approach to medicine. Springer, Berlin
105. Haller J, Toth M, De Boer SF (2006) Patterns of violent aggression-induced brain c-fos expression in male mice selected for aggressiveness. Physiol Behav 88:173–182
106. Van der Vegt BJ, Lieuwes N, van de Wall EH, Kato K, Moya-Allbiol L, Martinez-Sanchis S, De Boer SF, Koolhaas JM (2003) Activation of serotonergic neurotransmission during the performance of aggressive behavior in rats. Behav Neurosci 117:667–674
107. Veenema AH, Neumann ID (2007) Neurobiological mechanisms of aggression and stress coping: a comparative study in mouse and rat selection lines. Brain Behav Evol 70:274–285
108. Mongeau R, Miller GA, Chiang E, Anderson DJ (2003) Neural correlates of competing fear behaviors evoked by an innately aversive stimulus. J Neurosci 23:3855–3868
109. Gozzi A, Jain A, Giovanelli A, Bertollini C, Crestan V, Schwarz AJ, Tsetsenis T, Ragozzino D, Gross CT, Bifone A (2010) A neural switch for active and passive fear. Neuron 67: 656–666
110. De Boer SF, Koolhaas JM (2005) 5-HT1A and 5-HT1B receptor agonists and aggression: a pharmacological challenge of the serotonin deficiency hypothesis. Eur J Pharmacol 526:125–139
111. Valentino RJ, Lucki I, Van Bockstaele EJ (2010) Corticotropin-releasing factor in the dorsal raphe nucleus: linking stress coping and addiction. Brain Res 1314:29–37
112. Waselus M, Valentino RJ, Van Bockastaele EJ (2011) Collateralized dorsal raphe nucleus projections: a mechanism for the integration of diverse functions during stress. J Chem Neuroanat 41:266–280
113. Hammack SE, Richey KJ, Schmid MJ, LoPresti ML, Watkins LR, Maier SF (2002) The role of corticotropin-releasing hormone in the dorsal raphe nucleus in mediating the behavioral consequences of uncontrollable stress. J Neurosci 22:1020–1026
114. Hammack SE, Schmid MJ, LoPresti ML, Der-Avakian A, Pellymounter MA, Foster AC, Watkins LR, Maier SF (2003) Corticotropin releasing hormone type 2 receptors in the dorsal raphe nucleus mediate the behavioral consequences of uncontrollable stress. J Neurosci 23:1019–1025
115. Kirby LG, Rice KC, Valentino RJ (2000) Effects of corticotropin-releasing factor on neuronal activity in the serotonergic dorsal raphe nucleus. Neuropsychopharmacology 22:148–162
116. Pernar L, Curtis AL, Vale WW, Rivier JE, Valentino RJ (2004) Selective activation of corticotropin-releasing factor-2 receptors on neurochemically identified neurons in the rat dorsal raphe nucleus reveals dual actions. J Neurosci 24:1305–1311
117. Amat J, Tamblyn JP, Paul ED, Bland ST, Amat P, Foster AC, Watkins LR, Maier SF (2004) Microinjection of urocortin 2 into the dorsal raphe nucleus activates serotonergic neurons and increases extracellular serotonin in the basolateral amygdala. Neuroscience 129: 509–519
118. Lowry CA, Rodda JE, Lightman SL, Ingram CD (2000) Corticotropin-releasing factor increases in vitro firing rates of serotonergic neurons in the rat dorsal raphe nucleus: evidence for activation of a topographically organized mesolimbocortical serotonergic system. J Neurosci 20:7728–7736
119. Lukkes JL, Forster GL, Renner KJ, Summers CH (2008) Corticotropin-releasing factor 1 and 2 receptors in the dorsal raphé differentially affect serotonin release in the nucleus accumbens. Eur J Pharmacol 578:185–193
120. Staub DR, Evans AK, Lowry CA (2006) Evidence supporting a role for corticotropin-releasing factor type 2 (CRF_2) receptors in the regulation of subpopulations of serotonergic neurons. Brain Res 1070:77–89

121. Valentino RJ, Liouterman L, van Bockastaele EJ (2001) Evidence for regional heterogeneity in corticotropin-releasing factor interactions in the dorsal raphe nucleus. J Comp Neurol 435:450–463
122. Waselus M, Nazzaro C, Valentino RJ, Van Bockastaele EJ (2009) Stress-induced redistribution of corticotropin-releasing factor receptor subtypes in the dorsal raphe nucleus. Biol Psychiatry 66:76–83

Chapter 2
Serotonin in the Nervous System of Vertebrates

2.1 Synthesis and Metabolism of Serotonin

Serotonin is synthesized in a two-step reaction, from tryptophan to 5-hydroxytryptophan (5-HTP) and then to 5-hydroxytryptamine (serotonin, 5-HT). The latter step is catalyzed by aromatic amino acid decarboxylase (AADC; EC 4.1.1.28), while the first step is catalyzed by tryptophan hydroxylase (TPH; EC 1.14.16.4), the rate-limiting enzyme in serotonin synthesis [1] (Fig. 2.1). TPH is present in the brain mostly in its second isoform, TPH2 [2, 3]. The enzyme uses Fe^{2+} as a cofactor, and oxygen and tetrahydrobiopterin (BH4) as co-substrates. In vitro evidence suggests that the autoregulation of the firing rate of serotonergic neurons depends on sustained tryptophan hydroxylase activity [4]. Consistently with that hypothesis, depolarization or increases in extracellular calcium increases the affinity of TPH for tryptophan and tetrahydrobiopterin [5, 6].

BH4 concentrations in the central nervous system are not at saturation level for TPH [7], and therefore TPH exerts only a fraction of its maximum velocity (V_{max}). Microinjections of BH4 in the brain increase TPH activity [7, 8]. BH4 at physiological levels also enhances the inactivation of TPH by nitric oxide via attack on free protein thiol groups [9], another mechanism which could regulate serotonin synthesis.

The strict requirement for Fe^{2+} and the inhibition of TPH by Fe^{3+} means that cellular metabolism of iron is also critical for serotonin synthesis [10, 11]. This inhibition is dependent on the oxidation of H_2O_2 to O_2 and Fe^{2+} to Fe^{3+}, with the subsequent formation of reactive oxygen species [10]. Therefore, TPH activity is strictly dependent on the redox state of the cell.

In addition to the regulation of TPH activity by its co-factors and co-substrates, another important mechanism of regulation is phosphorylation. Phosphorylation by protein kinase A (PKA; EC 2.7.11.11) increases the V_{max} of TPH without changing its K_m [12]; similar results were obtained regarding phosphorylation by Ca^{2+}/calmodulin-dependent protein kinase II (CaMKII; EC 2.7.11.17) [13, 14].

Tryptophan hydroxylase is also activated by stress hormones. Psychological or physical stressors, as well as corticosteroid administration, increase TPH activity

C. Maximino, *Serotonin and Anxiety*, SpringerBriefs in Neuroscience,
DOI: 10.1007/978-1-4614-4048-2_2, © The Author(s) 2012

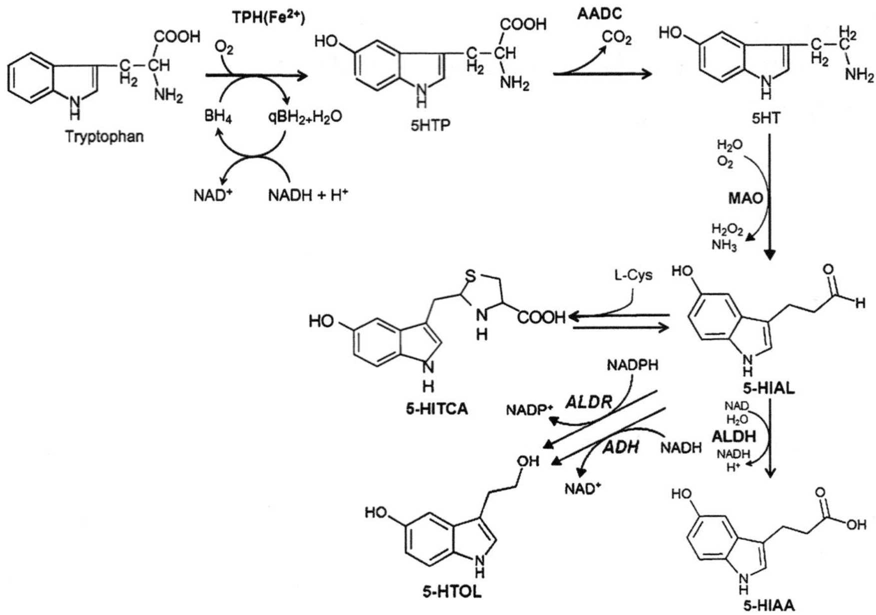

Fig. 2.1 Synthesis and degradation of serotonin. Tryptophan hydroxylase (*TPH*) is the rate-limiting enzyme in the synthesis of serotonin, converting tryptophan into 5-hydroxytryptophan (*5-HTP*). This intermediate step is followed by the conversion of 5-HTP into serotonin by amino acid decarboxylase (*AADC*). The principal enzyme in serotonin degradation is monoamine oxidase (*MAO*), which catalyzes the oxidative deamination of serotonin into 5-hydroxy-3-indolacetaldehyde (*5-HIAL*). This metabolite can follow three conversion pathways: in the presence of high L-cysteine (*L-Cys*) concentrations, 5-HIAL is transformed into 5-hydroxyindole thiazoladine carboxylic acid (*5-HITCA*). Aldehyde dehydrogenase (*ALDH*) catalyzes the conversion of 5-HIAL into 5-hydroxyindolacetic acid (*5-HIAA*). Chronic ethanol intake increases activity and/or concentrations of alcohol dehydrogenase (*ADH*) or aldehyde reductase (*ALDR*), which pushes the balance toward the conversion of 5-HIAL into 5-hydroxytryptophol (*5-HTOL*). Other abbreviations: *BH2* dihydrobiopterin, *BH4* tetrahydrobiopterin, *NAD+* nicotinamide adenine dinucleotide, *NADH* reduced NAD, *NADP* NAD phosphate, *NADPH* reduced NADP

in the brain [15–17]; in vitro, tryptophan hydroxylase activity is increased in the caudal part of the DRN by application of CRF [17]. This increase in activity is likely caused by alterations in gene expression; neonatal handling decreases, and maternal separation increases the expression of *tph*2 mRNA in caudal and ventrolateral parts of the DRN in rats which are exposed to social defeat as adults [18, 19]. In adult rats, chronic unpredictable stress increases *tph*2 mRNA expression in the caudal part of the DRN, but not in other raphe regions [20].

*Tph*2 is under the transcriptional control of three important developmental genes, *sim*1, *pet*1, and *lmx*1*b* [21–23]. In the case of the last two genes, adult deletion downregulates the expression of tryptophan hydroxylase 2 in the DRN [24, 25], suggesting that the environmental regulation of the expression of *tph*2 might be mediated by direct transcriptional activation by these developmental genes. Indeed, Pet-1 mRNA has been shown to be upregulated by estradiol treatment in ovariectomized rats [26], demonstrating that this gene is susceptible to extrinsic regulation.

Tryptophan hydroxylase inhibition by pharmacological means [through DL-*para*-chloropheynalainine (DL-pCPA)] has an anticonflict effect [27, 28], increases social interaction [29], and prevents learned helplessness after exposure to unpredictable stress [30]; conversely, panicogenic effects of DL-pCPA were observed in escape paradigms in which lever-press turns off dorsal periaqueductal gray area electrical stimulation [31, 32]. Deletion of *tph2* dramatically decreases brain 5-HT content and produces an anxiogenic effect in the marble burying and open-field tests [33]. Likewise, in BALB/c mice, a strain which is spontaneously more anxious and reactive than the C57BL/6 or 129Sv strains [34, 35], a single-nucleotide polymorphism at *tph2* leads to a decreased catalytic activity of the enzyme and, consequently, decreased serotonin levels in the forebrain [36]. The discrepancy between the behavioral effects of developmental and adult blockade of TPH2 activity suggests that serotonin has an important developmental role in modeling fear and anxiety circuits [37].

After reuptake from the synaptic cleft (see Sect. 2.2, below), serotonin is metabolized by monoamine oxidase (MAO; EC 1.4.3.4) [38], an enzyme that is located in the outer mitochondrial membrane [39]. This enzyme catalyzes the oxidative deamination of serotonin by converting it into 5-hydroxy-3-indolacetaldehyde (5-HIAL), which is further metabolized into 5-hydroxy-3-indolacetic acid (5-HIAA) by aldehyde dehydrogenase type 2 (ALDH2; EC 1.2.1.3) or to 5-hydroxytryptophol (5-HTOL) by aldehyde reductase (ALDR; EC 1.1.1.21) or alcohol dehydrogenase (ADH; EC 1.1.1.1). This latter metabolite is present in the brain at concentrations of only 1–5 % of 5-HIAA levels [40], but can be dramatically increased by ethanol intake [41]. Another minor metabolite is 5-hydroxyindole thiazoladine carboxylic acid (5-HITCA), which is formed from a condensation reaction between L-cysteine and 5-HIAL [42] (Fig. 2.1).

The chemical reaction catalyzed by MAO uses flavin adenine dinucleotide (FAD) as a redox cofactor [43]. Covalent attachment of FAD to cysteine residues is critical for enzymatic activity [44]. The first step in the reaction is reduction of FAD to $FADH_2$, coupled to the conversion of the amine group of serotonin to imine; a second step is the oxidation of $FADH_2$ with the conversion of oxygen into hydrogen peroxide, followed by the hydrolysis of the imine into the corresponding aldehyde and ammonia [45].

Monoamine oxidase is present in two isoforms in mammals, MAO A and MAO B, with the first having higher affinity for serotonin than MAO B [46]. Both isozymes are found in serotonergic neurons, but in different compartments: while MAO B is found in mitochondria at the soma, MAO A is found in mitochondria at the synaptic terminals [45]. This compartmentalization responds to the physiological requirements of serotonin synthesis and metabolism: 5-HT synthesis is much greater in the cell body than in axon terminals [47, 48]; if MAO A was to be localized in the soma, metabolism of serotonin would compete with the gradient-based mechanism of 5-HT vesicular uptake [45].

MAO A inhibition increases serotonin concentrations in the synaptic cleft while decreasing the firing rate of serotonergic neurons in the raphe [49, 50]. These effects are thought to underlie the antidepressant efficacy of MAO inhibitors. This class of drugs also has clinical efficacy in the treatment of panic and post-traumatic stress disorders, but not in generalized anxiety disorder (Table 1.1). Consistently, MAO A knockout mice do not show alterations in anxiety-like behavior in the plus-maze [51].

2.2 Transport of Serotonin: SERT and Uptake$_2$

The extracellular levels of serotonin must be tightly controlled by uptake into astrocytes and presynaptic neurons in order to fulfill the behavioral and physiological functions of serotonin. At least two mechanisms are responsible for this uptake. The most well known and well studied of these mechanisms is transport by a high-affinity, sodium- and chloride-dependent transporter, SERT (5-HTT, SLC6A4) [52, 53]. This protein is highly concentrated in the raphe nuclei and in the prefrontal cortex, as well as in thalamocortical afferents [54–56]. A physical interaction between the serotonin transporter and type 1 nitric oxide synthase (L-arginine, NADPH oxygen oxidoreductase 1; NOS-1; EC 1.14.13.39) has been observed; this interaction decreases SERT membrane trafficking, and in human embryonic kidney HEK293 cells transfected with SERT and NOS-1 serotonin leads to cyclic guanosine monophosphate (cGMP) formation through a transport-dependent mechanism. Increased NO production in SERT/NOS-1-transfected HEK293 cells triggered by a calcium ionophore or by nitric oxide donors does not affect serotonin reuptake, but NOS-1-deficient mice show increased 5-HT uptake [57].

Likewise, the isoform Iα of cGMP-dependent protein kinase (PKGIα; EC 2.7.11.12) colocalizes with SERT in both intracellular and cell surface domains of RN46A cells, an immortalized lineage of rat serotonergic raphe neurons, and is co-immunoprecipitated with SERT when stably transfected in HeLa cells [58]. Different from NOS-1, however, the interaction with PKGIα increases surface expression of SERT [59]. In RN46A cells and in the basophilic leukemia cell line RBL-2H3, the administration of 8-bromo-cGMP, a PKG activator, stimulates SERT surface trafficking [60] and serotonin uptake [61], an effect which is blocked by pretreatment with a membrane-permeant kinase inhibitor [58] and with a p38 mitogen-activated protein kinase (p38 MAPK; EC 2.7.12.2) inhibitor or a protein phosphatase 2A (PP2A; EC 3.1.3.16) inhibitor [62]. Likewise, treatment with a p38 MAPK inhibitor decreases the interaction between SERT and the PP2A catalytic subunit [63], suggesting that PP2A is downstream of p38, which is downstream to PKG. SERT activation by PKG is also blocked by mutation of a threonine residue near the cytoplasmic end of the fifth transmembrane helix [64], suggesting that the PKG-dependent, p38-independent phosphorylation of this

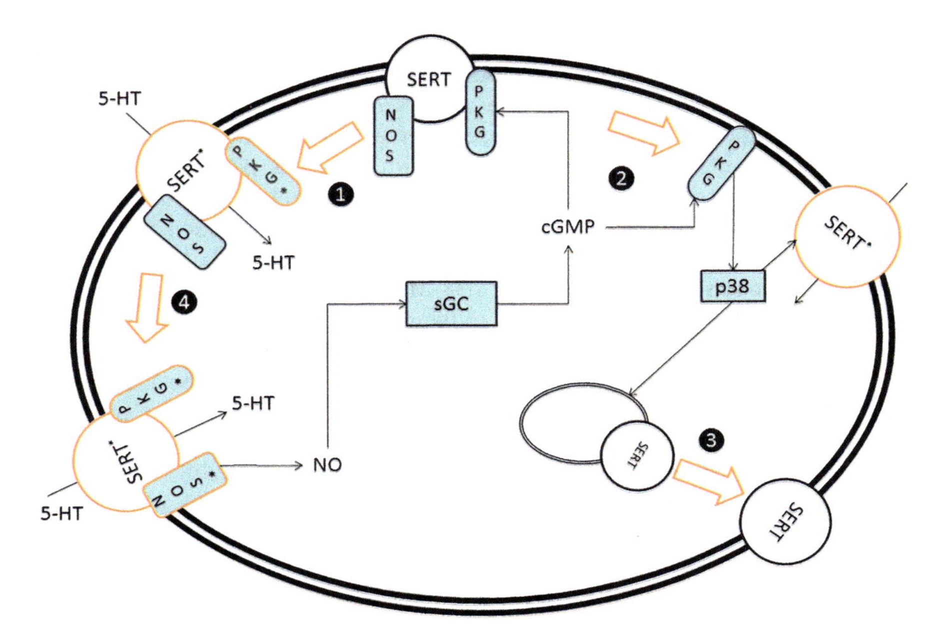

Fig. 2.2 Mechanisms of SERT regulation by the NO-cGMP pathway. *1* cGMP can activate PKGIα, which is linked to SERT via myosin; this mechanism stimulates serotonin uptake. *2* Alternatively, cGMP can activate PKG which phosphorylates p38 MAPK, leading to transport stimulation or *3* SERT membrane trafficking. SERT activation can also lead to activation of NOS-1 and subsequent production of nitric oxide *4*. Abbreviations: *NOS* nitric oxide synthase, *p38* p38 mitogen-activated protein kinase, *PKG* protein kinase G, *SERT* serotonin transporter, *sGC* soluble guanylate cyclase

residue is also necessary for this effect. Nonetheless, expression of PKGIα or PKGI*β* along with SERT in HeLa cells increases surface expression of the transporter even when the transfected PKG was mutated so as to eliminate its catalytic activity [59]. Transfection of a mutant form of SERT which lacks sites for *N*-glycosylation into Chinese hamster ovary (CHO) cells impair the association between myosin (a PKG-anchoring protein) and the transporter as well as the stimulation of uptake function by 8-bromo-cGMP, suggesting that the phosphorylation of SERT by PKG is dependent on the interaction of both proteins with myosin and on SERT being in a favorable conformation that is promoted by sialylated *N*-glycans [65] (Fig. 2.2).

Interestingly, p38 MAPK also seems to regulate the interaction between SERT and syntaxin 1A (Syn1A), a protein which determines the stoichiometry of 5-HT uptake by SERT [63]; when both proteins interact, 5-HT uptake is electroneutral, without 5-HT-induced ionic currents or 5-HT-independent sodium currents. When this interaction is broken, 5-HT uptake is electrogenic due to the appearance of a voltage-dependent influx of 12 sodium ions per 5-HT molecule (at −80 mV), which can lead to cell depolarization [66]. CaMKII inhibitors shift the affinity of

SERT for Syn1A, leading to the emergence of these SERT-mediated currents [67]. Importantly, p38 MAPK inhibitors produce the same effect [63], suggesting that p38 also regulates the conducting state of the serotonin transporter.

In spite of the fact that most of these observations were made in transfection systems, the regulation of SERT by the PKG-p38 MAPK pathway is of physiological significance. A non-synonymous single-nucleotide polymorphism in human SERT (I425 V) has high rates of serotonin uptake in relation to wild-type transporters, an effect which is associated with insensitivity to treatment with nitric oxide donors [68]. Isoform α of p38 MAPK is widely distributed in tryptophan hydroxylase 2-expressing cells in the mouse DRN [69], and social defeat stress leads to p38 phosphorylation in this region [70]. Adenoviral deletion of p38α in the DRN abolishes stress-induced reinstatement of cocaine seeking behavior, and blocks the acquisition of conditioned place avoidance of a context which has been paired with a κ-opioid receptor agonist, effects which are also observed with conditional deletion of p38α in SERT-expressing cells in the whole CNS but not in astrocytes. Also, whole-brain synaptosomes of mice injected with a κ-opioid receptor agonist show increased rate of SERT-specific serotonin uptake than synaptosomes of control animals; this effect, however, is not observed in the synaptosomes of mice with conditional deletion of p38α in SERT-expressing cells [70].

SERT is the target of serotonin selective reuptake inhibitors (SSRIs), and acute administration increases serotonin release preferentially in the raphe nuclei [71–74]. Genetic ablation of SERT results in elevated anxiety-like behavior in adult mice, an effect which is modulated by differences in genetic background [75] and is accompanied by fivefold increases in extracellular 5-HT concentrations [76], desensitization and downregulation of 5-HT$_{1A}$Rs in the raphe, hypothalamus, septum, and amygdala [77], 5-HT$_{1A}$R-dependent reduction in spontaneous firing rates in 5-HTergic neurons in the DRN [78], downregulation of 5-HT$_{1B}$Rs in serotonergic axon terminals [79], and disorganization of vibrissae-stimulated barrel fields in the somatosensory cortex [80]. Early blockade of SERT by chronic SSRI treatment mimics the effect of knocking out the transporter [81]. SERT overexpression, on the other hand, results in decreased anxiety [82].

Wistar-Zagreb 5-HT rats selected for extreme values of peripheral (platelet) transporter activity [83, 84] show discordant brain 5-HT turnover and extracellular concentrations, for animals with high peripheral SERT activity (high WZ-5HT) presenting normal serotonin levels in the brain, slightly lower [^{3}H]citalopram binding sites and increased serotonin release after potassium chloride or citalopram infusion in a microdialysis probe on the ventral hippocampus [85]. Thus, high WZ-5HT rats are not exactly hyperserotonergic, in the sense that their baseline serotonergic "tone" is not different from control populations or from low WZ-5HT rats; nonetheless, their serotonergic system is "hyperresponsive" to stimulation [84]. Interestingly, there are no differences between both lines of Wistar-Zagreb rats and control populations in terms of 5-HT$_{1A}$R, 5-HT$_{1B}$R, or 5-HT$_{2A}$R mRNA or protein expression, nor of binding sites or affinity to radioligands, in the hippocampus, striatum, or cortex [84, 86], in opposition to the compensatory changes observed in SERT knockout mice [77, 79]. High WZ-5HT rats show increased anxiety in the elevated plus-maze and social interaction tests, as well as diminished exploratory activity in the elevated zero-maze plus-maze and in the holeboard [87], as well as increased freezing to context [84], resembling SERT knockout animals [75].

These WZ-5HT animals have been proposed as a rodent model of human serotonergic dysfunctions associated with SERT genetic polymorphisms [84]. In humans (as well as in rhesus macaques [88, 89]), an important genetic polymorphism rests in a promoter region of the *slc6a4* gene. This polymorphic region (5-HTTLPR) consists of a 44 base-pair insertion/deletion event expressed as a short (*s*) or long (*l*) allele [90]; the long allele, with 16 repeat elements, leads to increased SERT expression than the short allele, with 14 repeat elements [90]. In immortalized lymphoblast B cell lines which express SERT in an allele-specific fashion, cells with the *l/l* genotype show higher 5-HT uptake and SERT protein expression than cells with the *l/s* or *s/s* genotypes [91], and brain-derived neurotrophic factor (BDNF) [92] and the adenylate cyclase activator forskolin [91] decrease 5-HT uptake in *l/l* lymphoblasts. The absence of an effect of these treatments on 5-HT uptake in *l/s* or *s/s* genotype lymphoblasts reinforces the idea of a decreased regulatory capacity of SERT in these genotypes.

Importantly, the presence of the short allele of the 5-HTTLPR polymorphism has been associated with anxiety symptoms in adult humans [91], and there are evidences of associations between this polymorphism and certain physiological and behavioral endophenotypes of anxiety. Carriers of the short allele exhibit increased startle responses to unexpected loud sounds [93], risk aversion [94–97], amygdala activity to fearful faces and to novel neutral stimuli [98–100], and functional coupling between amygdala and ventromedial prefrontal cortex when viewing positive or negative emotionally charged images [101, 102].

The rhesus macaque 5-HTTLPR also contains a polymorphism consisting on a 21 base-pair insertion/deletion event expressed as short or long alleles [89]. The presence of two copies of the *s* allele leads to ~50 % less serotonin transporters and is associated with lower 5-HIAA levels [103]. Importantly, it also produces behavior that is indicative of higher anxiety levels [104]. Infant macaques carrying the *s* allele who underwent short maternal separation periods showed higher levels of circulating ACTH during separation [105]. ACTH responses were higher in female *l/s* than *l/l* macaques only if they were allowed sensory contact with other subjects in addition to the maternal separation event, [106]. By contrast, *s* carrier males showed increased ACTH secretion regardless of rearing background [105]. Likewise, *s* carrier macaques also show greater sensitivity to intravenous injections of alcohol when adolescent [105], and female *s* carrier macaques spontaneously consume more alcohol when exposed to maternal separation [107].

In addition to the high-affinity sodium-dependent SERT, a low-affinity sodium-independent transporter has been described, a system called uptake$_2$. Uptake$_2$ is a high-capacity, low-affinity transport system for monoamines that is acutely inhibited by corticosteroids [108, 109]. Two proteins mediate uptake$_2$ in the CNS, the plasma membrane monoamine transporter (PMAT) and the organic cation transporter 3 (OCT3) [110]. Human OCT3 is the major uptake$_2$ transporter of histamine, epinephrine, and norepinephrine, while human PMAT is the major uptake$_2$ transporter for serotonin and dopamine [110]. Nonetheless, the affinity of corticosteroids to OCT3 is much higher than to PMAT [111, 112]. While PMAT is highly expressed in the mouse cortex, olfactory tubercle and hippocampus [113], OCT3, is expressed in highest densities in the colliculi islands of Calleja, subiculum, lateral septum, lateral and dorsomedial hypothalamic nuclei, and retrosplenial cortex [114]. Despite the low selectivity for serotonin, OCT3 has been implicated in 5-HT transport, as SERT-deficient mice show upregulation of OCT3 expression in the hippocampus [115], and inhibition of OCT3 in the rat dorsomedial hypothalamus dramatically increases extracellular serotonin levels in this structure [116]. The effects of OCT3 deletion on anxiety-like behavior are not contradictory, and probably dependent on the targeting mechanism and on background genotype. For example, in mice from a 129/Ola-FVB/N genetic background (25 %:75 %), deletion of exons 1 and 2 produce animals with increased anxiety-like behavior in the elevated Y-maze and open-field [117]; deletion of exons 6 and 7 in a C57BL/6 genetic background leads to decreased anxiety-like behavior in the elevated plus-maze and open-field [118].

2.3 Serotonin Receptors

Up to 14 distinct human serotonin receptors have been cloned [119], and most of these receptors present post-transcriptional modifications [120]. While the 5-HT$_3$ receptor is ionotropic, all other receptor classes (5-HT$_1$-5-HT$_7$) are metabotropic. Canonical transduction pathways for these receptors have been described as positive (5- 5-HT$_4$, 5-HT$_6$, 5-HT$_7$) or negative (5-HT$_1$, 5-HT$_5$) coupling to adenylyl cyclase and positive coupling to phosphatidyl inositide turnover (5-HT$_2$) [121]. Non-canonical transduction pathways have also been described, including nitric oxide synthase and phospholipase A$_2$ activation (Table 2.1) [122, 123].

Specific serotonin receptors have been linked to anxiety by pharmacological [124] and genetic [125] experiments. Among those, evidence for important roles in anxiety has been gathered for the 5-HT$_{1A}$ and 5-HT$_{2C}$ receptors. The main difficulty in evaluating these results is that these receptors are located both presynaptically at the raphe nuclei and postsynaptically at its targets. In the raphe, 5-HT$_{1A}$ receptors are somatodendritic, while 5-HT$_{1B}$ receptors show a preterminal axonal localization [126, 127]. 5-HT$_{1A}$ receptors are also enriched in limbic regions of the brain, while 5-HT$_{1B}$ receptors are located mainly on the basal ganglia [128, 129].

Table 2.1 Major signaling pathways of G-protein coupled 5-HT receptors

Receptor	Canonical transduction pathway	Secondary transduction pathways
5-HT$_{1A}$	AC inhibition	MAPK cascade activation
	GIRK modulation	PLC activation
	HVA inhibition	PLA2 activation/inhibition
		NOS-1 activation/inhibition
		Jak cascade activation
5-HT$_{1B}$	AC inhibition	MAPK cascade activation
		PLC activation
		NOS-1 activation
		GIRK modulation
5-HT$_{1D}$	AC inhibition	HVA inhibition
		GIRK activation
5-HT$_{1E}$	AC inhibition	AC activation (weak)
5-HT$_{1F}$	AC inhibition	PLC activation
5-HT$_{2A}$	PLC activation	PLA_2 activation
		Jak cascade activation
		GPase activation (astroglia)
		MMP activation (astroglia)
5-HT$_{2B}$	PLC activation	NOS-2 activation
		MAPK cascade activation
		PLA_2 activation
5-HT$_{2C}$	PLC activation	Na^+/Ca^{2+} exchanger activation
		GIRK inhibition
		PLA_2 activation
5-HT$_4$	AC activation	
5-HT$_{5A}$	AC inhibition	PLC activation (transient)
		GIRK activation
		ADP-ribosyl cyclase inhibition
5-HT$_{5B}$	Unknown	
5-HT$_6$	AC5 activation	
5-HT$_7$	AC5 activation	MAPK cascade activation
		cAMP-GEF cascade activation

Abbreviations: *AC* adenylyl cyclase, *cAMP-GEF* cyclic adenosine monophosphate-binding guanine nucleotide exchange factor, *GIRK* G-protein coupled inward rectifying potassium channel, *GPase* glycogen phosphorylase, *HVA* high-voltage activated calcium channel, *NOS* inducible nitric oxide synthase, *MAPK* mitogen-activated protein kinase, *MMP* matrix metalloproteinase, *NOS* nitric oxide synthase, *PLA_2* phospholipase A_2, *PLC* phospholipase C

2.3.1 5-HT1A Receptors

Different transduction mechanisms mediate 5-HT$_{1A}$R activation in presynaptic (DRN) and postsynaptic sites: in the raphe, activation of 5-HT$_{1A}$Rs inhibits high-voltage activated currents (I_{HVA}), but do not couple to adenylyl cyclase [130]. Knockout animals in which the 5-HT$_{1A}$ receptor is absent from birth show increased anxiety-like behavior [131] that is rescued after normalizing expression of 5-HT_{1A}Rs in forebrain areas in adulthood [132]. Conversely, decreasing the density of 5-HT_{1A} autoreceptors in adulthood in normal animals does not affect baseline anxiety-like behavior but leads to exaggerated behavioral despair in the tail suspension and forced swim tests after prolonged stress [133]. When this suppression is made at birth and maintained throughout life, however, increased anxiety-like behavior is observed in the open-field and light/dark box tests, while behavioral despair remains normal [134]. It seems that the developmental maintenance of a serotonergic "tone", mediated by 5-HT_{1A} autoreceptors, has an important role in the organization of defensive behavior.

5-HT_{1A} receptor expression, activity, and/or trafficking are regulated by a plethora of mechanisms. The coupling of 5-HT_{1A} receptors to different transduction mechanisms depend on its interaction with $G\alpha_{i/o}$ or $G\beta\gamma$ proteins, with the latter being necessary for I_{HVA} inhibition [135]. Coupling to $G\alpha_{i/o}$ (and, therefore, to adenylyl cyclase) is tightly regulated by regulator of G protein signaling (RGS) proteins, which function as GTPase accelerating proteins (GAPs) that bind to and rapidly deactivate $G\alpha$ subunits [136]. Mice with a point mutation in $G\alpha_{i2}$ that renders this protein insensitive to RGS protein regulation show less neophobia and reduced anxiety in the elevated plus-maze and marble burying test. They also show a dramatically increased response to treatment with 8-OH-DPAT or fluvoxamine in the tail suspension test [137]. Interestingly, RGS4 expression in serotonergic neurons is dependent on the activity of the transcription factor *Sim*1, which is enriched in lateral wings of the dorsal raphe nucleus [22].

The 5-HT_{1A} gene promoter possesses a negative glucocorticoid response element (GRE) that is highly sensitive to binding of a mineralocorticoid-glucocorticoid receptor heterodimer, less sensitive to binding of the mineralocorticoid receptor (MR), and weakly sensitive to binding of the glucocorticoid receptor (GR) [138]. Additionally, activation of GRs represses an NF-κB-induced 5-HT_{1A} promoter activity [139]. In rats, adrenalectomy is followed by a rapid upregulation of 5-HT_{1A} mRNA expression and 5-HT_{1A}R binding sites in the septo-hippocampal system, but not in the raphe [140–144]; this effect is reversed by treatment with low doses of corticosterone which preferentially bind MRs [140, 145–147]. Since both types of receptors are expressed in the septo-hippocampal system, while only GR is expressed in the DRN [139, 148] this affinity profile of the negative GRE in the 5-HT_{1A}R promoter could explain the lack of effect of manipulations of corticosteroid levels on autoreceptors in the raphe. In contrast, local activation of GRs in the DRN, as well as chronic mild stress, decreases the potency of 5-HT_{1A} receptor agonists to inhibit cell firing [144, 149–151]. These effects are not observed with acute stressors, however [152, 153], suggesting that corticosteroid effects on the raphe are due to a desensitization of 5-HT_{1A} autoreceptors [154].

The 5-HT_{1A} receptor gene is also negatively controlled by two calcium-dependent transcription factors, Freud-1 and Freud-2 (Five repressor element under dual binding protein) [155, 156]. In raphe cells, binding of Freud-1 in the dual repressor element (DRE) in the 5-HT_{1A} promoter region is sufficient to repress transcription; however, in postsynaptic cells, a second protein complex

(Freud-2) is necessary to reach the same effect [138, 156]. After chronic restraint stress, Freud-1 levels are reduced in the prefrontal cortex of male Sprague–Dawley rats, upregulating 5-HT_{1A}R mRNA expression; however, 5-HT_{1A}R protein levels are downregulated, suggesting post-transcriptional receptor downregulation [157]. Conversely, social defeat downregulates 5-HT_{1A} mRNA levels in the prefrontal cortex of Wistar rats, but no effect is seen in Freud-1 mRNA levels [158].

2.3.2 5-HT_{1B} Receptors

The primarily preterminal axonal localization of 5-HT_{1B} autoreceptors suggests that they may also have a role in the control of serotonergic "tone" [127]. Nonetheless, the effects of knocking down 5-HT_{1B}Rs are opposite to those observed after genetic manipulation of 5-HT_{1A}Rs. 5-HT_{1B}R knockout mice show increased exploratory activity in an open field that disappears after multiple expositions [159]; 5-HT_{1A}Rs, on the other hand, show decreased exploratory activity in the open field [131]. Moreover, while 5-HT_{1A} knockouts show increased anxiety in an elevated plus-maze and in the novelty-suppressed feeding test, 5-HT_{1B} knockouts show decreased anxiety [131, 159]. 5-HT_{1B} knockouts show decreased serotonin levels in the nucleus accumbens, locus coeruleus, and spinal cord, but not in regions associated with anxiety- or fear-like behavior [160]. Increasing expression of 5-HT_{1B}Rs in the dorsal raphe nucleus increases anxiety-like behavior in the elevated plus-maze and open field following inescapable stress, but decreased it in the absence of prior stress [161]; a similar effect is seen in the fear-potentiated startle paradigm, where 5-HT_{1B}R-overexpressing animals show decreased potentiation in the absence of prior stress, and this effect is abolished by inescapable stress [162]. Specific overexpression in the caudal DRN decreases immobility in the Porsolt forced swim test and freezing in contextual fear conditioning [163].

In normal animals, selection for increased learned helplessness decreases the expression of 5-HT_{1B} mRNA in the dorsal raphe nucleus; an opposite phenomenon is observed in animals which have been separated from their mothers for 15 min/day between postnatal days 2–14, but not in animals which have been separated for 180 min/day for the same period [164]. 5-HT_{1B} receptors interact with an S100 EF hand protein, S100A10, which increases membrane trafficking for this receptive protein. In a mouse line which has been selected for readiness of establishment of learned helplessness (H/Rouen mice), S100A10 mRNA is greatly diminished in the forebrain, and treatment with imipramine or electroconvulsive therapy increases S100A10 mRNA in the forebrain of rats. Overexpression of S100A10 in African Green monkey kidney COS-7 cells co-expressing 5-HT_{1B} receptors increases the amount of these receptors at the cell surface and increases the efficacy of adenylyl cyclase coupling. When S100A10 is overexpressed in the brains of mice, deceased thigmotaxis is seen in the open field, and animals spend less time immobile in the tail suspension test. S100A10 knockout mice show decreased

5-HT_{1B} receptor binding and decreased G-protein coupling efficacy in the globus pallidus in response to anpirtoline, a 5-HT_{1B} agonist. Primary cortical cultures from S100A10 knockout mice also lose their ability to respond to serotonin or anpirtoline with decreased extracellular signal-regulated kinase (ERK) phosphorylation. Finally, in S100A10 knockout mice treated chronically with imipramine, anpirtoline does not decrease thigmotaxis in an open field [165]. Overall, these results point to an important modulation of 5-HT_{1B} receptor function by S100A10 that is associated with stress reactivity, but not anxiety-like behavior. These results are also consistent with a role for the 5-HT_{1B} receptor in stress reactivity.

2.3.3 5-HT_{2C} Receptors

5-HT_{2C} receptors are located in raphe GABAergic interneurons [166] as well as in the nucleus accumbens, olfactory tubercle, claustrum, septum, cingulate cortex, amygdala, hippocampus, periaqueductal gray, habenula, and entorhinal cortex [128]. These receptors have been implicated in anxiety-like behavior, as *htr2c*$^{-/-}$ mice show decreased anxiety-like behavior [167], and the mixed $5\text{-HT}_{1B}/5\text{-HT}_{2A}/5\text{-HT}_{2C}$ agonists mCPP and 1-(3-trifuloromethylphenyl)piperazine (TFMPP) have anxiogenic effects in human subjects with anxiety disorders [168–170] and in animal models [171]. The discriminative stimulus effects of mCPP are mediated by 5-HT_{2C} receptors, as other 5-HT_{2C} agonists fully substitute for mCPP on drug discrimination assays, and 5-HT_{2C} antagonists block this effect [172]; thus, it seems that the interoceptive cues of this drug is 5-HT_{2C}R-specific.

5-HT_{2C}Rs are the most "variable" among the serotonergic receptors, being associated with "collateral efficacy" or agonist-directed trafficking of receptor stimulus. In 5-HT_{2C} receptors transfected with CHO-derived cells, agonists activate two signal transduction pathways (phospholipase C-mediated inositol phosphate accumulation and phospholipase A_2-mediated arachidonic acid release) differentially ; some drugs, such as TFMPP (3-trifuloromethylphenyl-piperazine), preferentially activate the PLC pathway, while others, such as LSD (lysergid acid diethylamide), preferentially activate the PLA_2 pathway [173]. Serotonin itself has a similar efficacy at each pathway, but is more potent at producing a response for inositol phosphate accumulation [173]. Importantly, agonist-directed trafficking is dependent on post-transcriptional-edited variants. 5-HT_{2C} receptors undergo adenosine-to-inosine editing events at five sites located within the second intracellular loop [174]. Two important isoforms ($5\text{-HT}_{2C\text{-VSV}}$ and $5\text{-HT}_{2C\text{-VGV}}$) have been confirmed in the central nervous system [174, 175], and, in relation to non-edited isoforms exhibit decreased constitutive activity, agonist affinity, and internalization [175–179]. Importantly, in edited isoforms, no difference in the efficacies of TFMPP or bufotenin for the PLC and PLA_2 pathways is observed [180]. The full significance of these findings is yet to be assessed in vivo, but it is known that the amount of RNA editing in the medial prefrontal cortex and amygdala is positively associated with anxiety-like traits in rodents [181, 182].

References

1. Jéquier E, Lovenberg W, Sjoerdsma A (1967) Tryptophan hydroxylase inhibition: the mechanism by which *p*-chlorophenylalanine depletes rat brain serotonin. Mol Pharmacol 3:274–278
2. Walther DJ, Bader M (2003) A unique central tryptophan hydroxylase isoform. Biochem Pharmacol 66:1673–1680
3. Gutknecht L, Kriegenbaum C, Waider J, Schmitt A, Lesch K-P (2009) Spatio-temporal expression of tryptophan hydroxylase isoforms in murine and human brain: convergent data from *Tph2* knockout mice. Eur Neuropsychopharmacol 19:266–282
4. Evans AK, Reinders N, Ashford KA, Christie IN, Wakerley JB, Lowry CA (2008) Evidence for serotonin synthesis-dependent regulation of in vitro neuronal firing rates in the midbrain raphe complex. Eur J Pharmacol 590:136–149
5. Boadle-Biber MC (1978) Activation of tryptophan hydroxylase from central serotonergic neurons by calcium and depolarization. Biochem Pharmacol 27:1069–1079
6. Boadle-Biber MC, Johannessen JN, Narasimhachari N, Phan T-H (1983) Activation of tryptophan hydroxylase by stimulation of central serotonergic neurons. Biochem Pharmacol 32:185–188
7. Miwa S, Watanabe Y, Hayaishi O (1985) 6R-L-erythro-5,6,7,8-tetrahydrobiopterin as a regulator of dopamine and serotonin biosynthesis in the rat brain. Arch Biochem Biophys 239:234–241
8. Kapatos G, Katoh S, Kaufman S (1982) Biosynthesis of biopterin by rat brain. J Neurochem 39:1152–1162
9. Kuhn DM, Arthur RE Jr (1997) Inactivation of tryptophan hydroxylase by nitric oxide: enhancement by tetrahydrobiopterin. J Neurochem 68:1495–1502
10. Hasegawa H, Nakamura K (2010) Tryptophan hydroxylase and serotonin synthesis regulation. In: Müller C, Jacobs B (eds) Handbook of behavioral neurobiology of serotonin. Elsevier B. V, Amsterdam, pp 183–202
11. Hasegawa H, Ichiyama A (2005) Distinctive iron requirement of tryptophan 5-monooxygenase: TPH1 requires dissociable ferrous iron. Biochem Biophys Res Commun 338:277–284
12. Johansen PA, Jennings I, Cotton RGH, Kuhn DM (1996) Phosphorylation and activation of tryptophan hydroxylase by exogenous protein kinase A. J Neurochem 66:817–823
13. Yamauchi T, Fujisawa H (1983) Purification and characterization of the brain calmodulin-dependent protein kinase (kinase II), which is involved in the activation of tryptophan 5-monooxygenase. Eur J Biochem 132:15–21
14. Yamauchi T, Nakata H, Fujisawa H (1981) A new activator protein that activates tryptophan 5-monooxygenase and tyrosine 3-monooxygenase in the presence of $Ca2^{+}$-, calmodulin-dependent protein kinase. Purification and characterization. J Biol Chem 256:5404–5409
15. Azmitia C Jr, McEwen BS (1976) Early response of rat brain tryptophan hydroxylase activity to cycloheximide, puromycin and corticosterone. J Neurochem 27:773–778
16. Azmitia EC Jr, McEwen BS (1974) Adrenalcortical influence on rat brain tryptophan hydroxylase activity. Brain Res 78:291–302
17. Evans AK, Heerkens JLT, Lowry CA (2009) Acoustic stimulation in vivo and corticotropin-releasing factor in vitro increase tryptophan hydroxylase activity in the rat caudal dorsal raphe nucleus. Neurosci Lett 455:36–41
18. Gardner KL, Thrivikraman KV, Lightman SL, Plotsky PM, Lowry CA (2005) Early life experience alters behavior during social defeat: focus on serotonergic systems. Neuroscience 136:181–191
19. Gardner KL, Hale MW, Oldfield S, Lightman SL, Plotsky PM, Lowry CA (2009) Adverse experience during early life and adulthood interact to elevate *tph2* mRNA expression in serotonergic neurons within the dorsal raphe nucleus. Neuroscience 163:991–1001

20. McEuen JG, Beck SG, Bale TL (2008) Failure to mount adaptive responses to stress results in dysregulation and cell death in the midbrain raphe. J Neurosci 28:8169–8177
21. Hendricks T, Francis N, Fyodorov D, Deneris ES (1999) The ETS domain factor Pet1 is an early and precise marker of central serotonin neurons and interacts with a conserved element in serotonergic genes. J Neurosci 19:10348–10356
22. Osterberg N, Wiehle M, Oehlke O, Heidrich S, Xu C, Fan C-M, Krieglstein K, Roussa E (2011) Sim1 is a novel regulator in the differentiation of mouse dorsal raphe serotonergic neurons. PLoS ONE 6:e19239
23. Cheng L, Chen C-L, Luo P, Tan M, Qiu M, Johnson R, Ma Q (2003) *Lmx1b*, *Pet*-1, and *Nkx2.2* coordinately specify serotonergic neurotransmitter phenotype. J Neurosci 23:9961–9967
24. Song N-N, Xiu J-B, Huang Y, Chen J-Y, Zhang L, Gutknecht L, Lesch K-P, Li H, Ding Y (2011) Adult raphe-specific deletion of *Lmx1b* leads to central serotonin deficiency. PLoS ONE 6:e15998
25. Liu C, Maejima T, Wyler SC, Casadesus G, Herlitze S, Deneris ES (2010) *Pet*-1 is required across different stages of life to regulate serotonergic function. Nat Neurosci 13:1190–1198
26. Rivera HM, Oberbeck DR, Kwon B, Houpt TA, Eckel LA (2009) Estradiol increases Pet-1 and serotonin transporter mRNA in the midbrain raphe nuclei of ovariectomized rats. Brain Res 1259:51–58
27. Söderpalm B, Engel JA (1989) Does the PCPA induced anticonflict effect involve activation of the GABAA/benzodiazepine chloride ionophore receptor complex? J Neural Transm 76:145–153
28. Robichaud RC, Sledge KL (1969) The effects of *p*-chlorophenylalanine on experimentally induced conflict in the rat. Life Sci 8:965–969
29. File SE, Hyde JRG (1977) The effects of *p*-chlorophenylalanine and ethanolamine-*O*-sulphate in an animal test of anxiety. J Pharm Pharmacol 29:735–738
30. Edwards E, Johnson J, Anderson D, Turano P, Henn FA (1986) Neurochemical and behavioral consequences of mild, uncontrollable shock: effects of PCPA. Pharmacol Biochem Behav 25:415–421
31. Kiser RS Jr, Lebovitz RM (1975) Monoaminergic mechanisms in aversive brain stimulation. Physiol Behav 15:47–53
32. Kiser RS, Lebovitz RM, German DC (1978) Anatomic and pharmacologic differences between two types of aversive midbrain stimulation. Brain Res 155:331–342
33. Savelieva KV, Zhao S, Pogorelov VM, Rajan I, Yang Q, Cullinan E, Lanthorn TH (2008) Genetic disruption of both tryptophan hydroxylase genes dramatically reduces serotonin and affects behavior in models sensitive to antidepressants. PLoS ONE 3:e3301
34. Stiedl O, Radulovic J, Lohmann R, Birkenfeld K, Palve M, Kammermeir J, Sananbenesi F, Spiess J (1999) Strain and substrain differences in context- and tone-dependent fear conditioning of inbred mice. Behav Brain Res 104:1–12
35. Yilmazer-Hanke DM, Roskoden T, Zilles K, Schwegler H (2003) Anxiety-related behavior and densities of glutamate, $GABA_A$, acetylcholine and serotonin receptors in the amygdala of seven inbred mouse strains. Behav Brain Res 145:145–159
36. Zhang X, Beaulieu J-M, Sotnikova TD, Gainetdinov RR, Caron MG (2004) Tryptophan hydroxylase-2 controls brain serotonin synthesis. Science 305:217
37. Azmitia EC (1999) Serotonin neurons, neuroplasticity, and homeostasis of neural tissue. Neuropsychopharmacology 21:33S–45S
38. Hare ML (1928) Tyramine oxidase: a new enzyme system in liver. Biochem J 22:968–979
39. Cotzias G, Dole V (1951) Metabolism of amines. II. Mitochondrial localization of monoamine oxidase. Proc Soc Exp Biol Med 78:157–160
40. Beck O, Borg S, Edman G, Fyro B, Oxenstierna G, Sedvall G (1984) 5-hydroxytryptophol in human cerebrospinal fluid: conjugation, concentration gradient, relationship to 5-hydroxyindoleacetic acid, and influence of hereditary factors. J Neurochem 43:58–61
41. Helander A, Beck O, Jacobsson G, Lowenmo C, Wikstrom T (1993) Time course of ethanol-induced changes in serotonin metabolism. Life Sci 53:847–855

42. Squires LN, Jakubowski JA, Stuart JN, Rubakhin SS, Hatcher NG, Kim W-S, Chen K, Shih JC, Seif I, Sweedler JV (2006) Serotonin catabolism and the formation and fate of 5-hydroxyindole thiazolidine carboxylic acid. J Biol Chem 281:13463–13470
43. Edmonson DE, Bhattacharyya AK, Walker MC (1993) Spectral and kinetic studies of imine product formation in the oxidation of p-(N,N-dimethylamino)benzylamine analogues by monoamine oxidase B. Biochemistry 32:5196–5202
44. Wu HF, Chen K, Shih JC (1993) Site-directed mutagenesis of monoamine oxidase A and B: role of cysteines. Mol Pharmacol 43:888–893
45. Bortolato M, Chen K, Shih JC (2010) The degradation of serotonin: role of MAO. In: Müller C, Jacobs B (eds) Handbook of behavioral neurobiology of serotonin. Elsevier B. V, Amsterdam, pp 203–218
46. Johnston JP (1968) Some observations upon a new inhibitor of monoamine oxidase in brain tissue. Biochem Pharmacol 17:1285–1297
47. Daszuta A, Hery F, Faudon M (1984) In vitro 3H-serotonin (5-HT) synthesis and release in BALBc and C57BL mice. I. Terminal areas. Brain Res Bull 12:559–563
48. Daszuta A, Faudon M, Hery F (1984) In vitro 3H-serotonin (5-HT) synthesis and release in BALBc and C57BL mice. II. Cell body areas. Brain Res Bull 12:565–570
49. Aghajanian GK, Graham AW, Sheard MH (1970) Serotonin-containing neurons in brain: depression of firing by monoamine oxidase inhibitors. Science 169:1100–1102
50. Sharp T, Gartside SE, Umbers V (1997) Effects of co-administration of a monoamine oxidase inhibitor and a 5-HT1A receptor antagonist on 5-hydroxytryptamine cell firing and release. Eur J Pharmacol 320:15–19
51. Popova NK, Skrinskaya YA, Amstislavskaya TG, Vishnivestkaya GB, Seif I, de Meider E (2001) Behavioral characteristics of mice with genetic knockout of monoamine oxidase type A. Neurosci Behav Physiol 31:597–602
52. Tripp A, Sibille E (2010) SERT models of emotional dysregulation. In: Kalueff AV, LaPorte JK (eds) Experimental models in serotonin transporter research. Cambridge University Press, New York, pp 105–134
53. Adell A, Celada P, Abellán MT, Artigas F (2002) Origin and functional role of the extracellular serotonin in the midbrain raphe nuclei. Brain Res Rev 39:154–180
54. Descarries L, Watkins KC, Garcia S, Beaudet A (1982) The serotonin neurons in nucleus raphe dorsalis of adult rats. J Comp Neurol 207:239–254
55. Fu W, Le Maître E, Fabre V, Bernard J-F, Xu Z-QD, Hökfelt T (2010) Chemical neuroanatomy of the dorsal raphe nucleus and adjacent structures of the mouse brain. J Comp Neurol 518:3464–3494
56. Qian Y, Melikian HE, Rye DB, Levey AI, Blakely RD (1995) Identification and characterization of antidepressant-sensitive serotonin transporter proteins using site-specific antibodies. J Neurosci 15:1261–1274
57. Chanrion B, la Cour CM, Bertaso F, Lerner-Natoli M, Freissmuth M, Millan MJ, Bockaert J, Marin P (2007) Physical interaction between the serotonin transporter and neuronal nitric oxide synthase underlies reciprocal modulation of their activity. Proc Nat Acad Sci USA 104:8119–8124
58. Steiner JA, Carneiro AMD, Wright J, Matthies HJG, Prasad HC, Nicki CK, Dostmann WR, Buchanan CC, Corbin JD, Francis SH, Blakely RD (2009) cGMP-dependent protein kinase Iα associates with the antidepressant-sensitive serotonin transporter and dictates rapid modulation of serotonin uptake. Mol Brain 2:26
59. Zhang Y-W, Rudnick G (2011) Myristoylation of cGMP-dependent protein kinase dictates isoform specificity for serotonin transporter regulation. J Biol Chem 286:2461–2468
60. Zhu CB, Carneiro AMD, Dostmann WR, Hewlett WA, Blakely RD (2005) p38 MAPK activation elevates serotonin transport activity via a trafficking-independent, PP2A-dependent process. J Biol Chem 280:15649–15658
61. Zhu CB, Carneiro AM, Dostmann WR, Hewlett WA, Blakely RD (2005) p38 MAPK activation elevates serotonin transport activity via a trafficking-independent, protein phosphatase 2A-dependent process. J Biol Chem 280:15649–15658

62. Zhu CB, Hewlett WA, Feoktistov I, Biaggioni I, Blakely RD (2004) Adenosine receptor, protein kinase G, and p38 mitogen-activated protein kinase-dependent up-regulation of serotonin transporters involves both transporter trafficking and activation. Mol Pharmacol 65:1462–1474
63. Samuvel DJ, Jayanthi LD, Bhat NR, Ramamoorthy S (2005) A role for p38 mitogen-activated protein kinase in the regulation of the serotonin transporter: evidence for distinct cellular mechanisms involved in transporter surface expression. J Neurosci 25:29–41
64. Ramamoorthy S, Samuvel DJ, Buck ER, Rudnick G, Jayanthi LD (2007) Phosphorylation of threonine residue 276 is required for acute regulation of serotonin transporter by cyclic GMP. J Biol Chem 282:11639–11647
65. Ozaslan D, Wang S, Ahmed BA, Kocabas AM, McCastlain JC, Bene A, Kilic F (2003) Glycosyl modification facilitates homo- and hetero-oligomerization of the serotonin transporter. J Biol Chem 278:43991–44000
66. Quick MW (2003) Regulating the conducting states of a mammalian serotonin transporter. Neuron 40:537–549
67. Ciccone MA, Timmons M, Phillips A, Quick MW (2008) Calcium/calmodulin-dependent kinase II regulates the interaction between the serotonin transporter and syntaxin 1A. Neuropharmacology 55:763–770
68. Kilic F, Murphy DL, Rudnick G (2003) A human serotonin transporter mutation causes constitutive activation of transport activity. Mol Pharmacol 64:440–446
69. Land BB, Bruchas MR, Schattauer S, Giardino WJ, Aita M, Messinger D, Hnasko TS, Palmiter RD, Chavkin C (2009) Activation of the kappa opioid receptor in the dorsal raphe nucleus mediates the aversive effects of stress and reinstates drug seeking. Proc Nat Acad Sci USA 106:19168–19173
70. Bruchas MR, Schindler AG, Shankar H, Messinger DI, Miyatake M, Land BB, Lemos JC, Hagan CE, Neumaier JF, Quintana A, Palmiter RD, Chavkin C (2011) Selective p38α MAPK deletion in serotonergic neurons produces stress resilience in models of depression and addiction. Neuron 71:498–511
71. Bel N, Artigas F (1992) Fluvoxamine preferentially increases extracellular 5-hydroxytryptamine in the raphe nuclei: an in vivo microdialysis study. Eur J Pharmacol 229:101–103
72. Gartside SE, Umbers V, Hajós M, Sharp T (1995) Interaction between a selective 5-HT_{1A} receptor antagonist and an SSRI in vivo: effects on 5-HT cell firing and extracellular 5-HT. Br J Pharmacol 115:1064–1070
73. Hervás I, Artigas F (1998) Effect of fluoxetine on extracellular 5-hydroxytryptamine in rat brani. Role of 5-HT autoreceptors. Eur J Pharmacol 358:9–18
74. Malagié I, Trillat AC, Jacquot C, Gardier AM (1995) Effects of acute fluoxetine on extracellular serotonin levels in the raphe: an in vivo microdialysis study. Eur J Pharmacol 286:213–217
75. Kalueff AV, Olivier JDA, Nonkes LJP, Homberg JR (2010) Conserved role for the serotonin transporter gene in rat and mouse neurobehavioral endophenotypes. Neurosci Biobehav Rev 34:373–386
76. Bengel D, Murphy DL, Andrews AM, Wichems CH, Feltner D, Heils A, Mossner R, Westphal H, Lesch K-P (1998) Altered brain serotonin homeostasis and locomotor insensitivity to 3,4-methylenedioxymethamphetamine ("ecstasy") in serotonin transporter-deficient mice. Mol Pharmacol 53:649–655
77. Li Q, Wichems C, Heils A, Lesch K-P, Murphy DL (2000) Reduction in the density and expression, but not G-protein coupling, of serotonin receptors (5-HT1A) in 5-HT transporter knock-out mice: gender and brain region differences. J Neurosci 20:7888–7895
78. Gobbi G, Murphy DL, Lesch K-P, Blier P (2001) Modifications of the serotonergic system in mice lacking serotonin transporters: an in vivo electrophysiological study. J Pharmacol Exp Ther 296:987–995

79. Fabre V, Rioux A, Lesch K-P, Murphy DL, Lanfumey L, Hamon M, Marres MP (2000) Altered expression and coupling of the serotonin 5-HT1A and 5-HT1B receptors in knock-out mice lacking the 5-HT transporter. Eur J Neurosci 12:2299–2310
80. Persico AM, Revay RS, Mössner R, Conciatori M, Marino R, Baldi A, Cabib S, Pascucci T, Sora I, Uhl GR, Murphy DL, Lesch K-P, Keller F (2001) Barrel pattern formation in somatosensory cortical layer IV requires serotonin uptake by thalamocortical endings, while vesicular monoamine release is necessary for development of supragranular layers. J Neurosci 21:6862–6873
81. Ansorge MS, Zhou M, Lira A, Hen R, Gingrich JA (2004) Early-life blockade of the 5-HT transporter alters emotional behavior in adult mice. Science 306:879–881
82. Jennings KA, Loder MK, Sheward WJ, Pei Q, Deacon RMJ, Benson MA, Olverman HJ, Hastie ND, Harmar AJ, Shen S, Sharp T (2006) Increased expression of the 5-HT transporter confers a low-anxiety phenotype linked to decreased 5-HT transmission. J Neurosci 26:8955–8964
83. Cicin-Sain L, Froebe A, Bordukalo-Niksic T, Jernej B (2005) Serotonin transporter kinetics in rats selected for extreme values of platelet serotonin levels. Life Sci 77:452–461
84. Cicin-Sain L, Jernej B (2010) Wistar-Zagreb 5HT rats: a rodent model with constitutional upregulation/downregulation of serotonin transporter. In: Kalueff AV, LaPorte JI (eds) Experimental models in serotonin transporter research. Cambridge University Press, Cambridge, pp 214–243
85. Romero L, Jernej B, Bel N, Cicin-Sain L, Cortes R, Artigas F (1998) Basal and stimulated extracellular serotonin concentration in the brain of rats with altered serotonin uptake. Synapse 28:313–321
86. Bordukalo-Niksic T, Mokrovic G, Stefulj J, Zivin M, Jernej B, Cicin-Sain L (2010) 5-HT1A receptors and anxiety-like behaviours: studies in rats with constitutionally upregulated/downregulated serotonin transporter. Behav Brain Res 213:238–245
87. Hranilovic D, Cicin-Sain L, Bordukalo-Niksic T, Jernej B (2005) Rats with constitutionally upregulated/downregulated platelet 5HT transporter: differences in anxiety-related behavior. Behav Brain Res 165:271–277
88. Rogers J, Kaplan J, Garcia R, Shelledy W, Nair S, Cameron J (2006) Mapping of the serotonin transporter locus (SLC6A4) to rhesus chromosome 16 using genetic linkage. Cytogenet Genome Res 112:341A
89. Lesch K-P, Meyer J, Glatz K et al (1997) The 5 HT transporter gene linked polymorphic region (5-HTTLPR) in evolutionary perspective: alternative biallelic variation in rhesus monkeys. J Neural Transm 104:1259–1266
90. Heils A, Teufel A, Petri S et al (1996) Allelic variation of human serotonin transporter gene expression. J Neurochem 66:2621–2624
91. Lesch K-P, Bengel D, Heils A, Sabol SZ, Greenberg BD, Petri S, Benjamin J, Müller CR, Hamer DH, Murphy D (1996) Association of anxiety-related traits with a polymorphism in the serotonin transporter gene regulatory region. Science 274:1527–1531
92. Mössner R, Daniel S, Albert D, Heils A, Okladnova O, Schmitt A, Lesch K-P (2000) Serotonin transporter function is modulated by brain-derived neurotrophic factor (BDNF) but not nerve growth factor (NGF). Neurochem Int 36:197–202
93. Brocke B, Armbruster D, Muller J et al (2006) Serotonin transporter gene variation impacts innate fear processing: acoustic startle response and emotional startle. Mol Psychiatry 11:1106–1112
94. Homberg JR, Van den Bos R, Den Heijer E, Suer R, Cuppen E (2008) Serotonin transporter dosage modulates long-term decision-making in rat and human. Neuropharmacology 55:80–84
95. Crisan LG, Pana S, Vulturar R, Heilman RM, Szekely R, Druga B, Dragos N, Miu AC (2009) Genetic contributions of the serotonin transporter to social learning of fear and economic decision making. Soc Cognitive Affect Neurosci 4:399–408
96. Kuhnen CM, Chiao JY (2009) Genetic determinants of financial risk taking. PLoS ONE 4:e4362

97. He Q, Xue G, Chen C, Lu Z, Dong Q, Lei X, Ding N, Li J, Li H, Chen C, Li J, Moyzis RK, Bechara A (2010) Serotonin transporter gene-linked polymorphic region (5-HTTLPR) influences decision making under ambiguity and risk in a large Chinese sample. Neuropharmacology 59:518–526
98. Hariri AR, Drabant EM, Munoz KE et al (2005) A susceptibility gene for affective disorders and the response of the human amygdala. Arch Gen Psychiatry 62:146–152
99. Hariri AR, Mattay VS, Tessitore A et al (2002) Serotonin transporter genetic variation and the response of the human amygdala. Science 297:400–403
100. Heinz A, Smolka MN, Braus DF et al (2007) Serotonin transporter genotype (5-HTTLPR): effects of neutral and undefined conditions on amygdala activation. Biol Psychiatry 61:1011–1014
101. Heinz A, Braus DF, Smolka MN et al (2005) Amygdala-prefrontal coupling depends on a genetic variation in the serotonin transporter. Nat Neurosci 8:20–21
102. Pezawas L, Meyer-Lindenberg A, Drabant EM et al (2005) 5-HTTLPR polymorphism impacts human cingulate-amygdala interactions: a genetic susceptibility mechanism for depression. Nat Neurosci 8:828–834
103. Bennet AJ, Lesch K-P, Heils A et al (2002) Early experience and serotonin transporter gene variation interact to influence primate CNS function. Mol Psychiatry 7:118–122
104. Herman KN, Winslow JT, Suomi SJ (2010) Primate models in serotonin transporter research. In: Kalueff AV, LaPorte JI (eds) Experimental models in serotonin transporter research. Cambridge University Press, Cambridge, pp 288–307
105. Barr CS, Newman TK, Becker ML et al (2003) Serotonin transporter gene variation is associated with alcohol sensitivity in rhesus macaques exposed to early life stress. Alcohol Clin Exp Res 27:812–817
106. Barr CS, Newman TK, Schwandt M et al (2004) Sexual dichotomy of an interaction between early adversity and the serotonin transporter gene promoter variant in rhesus macaques. Proc Nat Acad Sci USA 101:12358–12363
107. Barr CS, Newman TK, Lindell S et al (2004) Interaction between serotonin transporter gene variation and rearing condition in alcohol preference and consumption in female primates. Arch Gen Psychiatry 61:1146–1152
108. Iversen LL, Salt PJ (1970) Inhibition of catecholamine uptake-2 by steroids in the isolated rat heart. Br J Pharmacol 40:528–530
109. Hill JE, Makky K, Shrestha L, Hillard CJ, Gasser PJ (2010) Natural and synthetic corticosteroids inhibit uptake$_2$-mediated transport in CNS neurons. Physiol Behav (in press)
110. Duan H, Wang J (2010) Selective transport of monoamine neurotransmitters by human plasma membrane membrane monoamine transporter and organic cation transporter 3. J Pharmacol Exp Ther 335:743–753
111. Gasser PJ, Lowry CA, Orchinik M (2006) Corticosterone-sensitive monoamine transport in the rat dorsomedial hypothalamus: potential role for organic cation transporter 3 in stress-induced modulation of monoaminergic neurotransmission. J Neurosci 26:8758–8766
112. Engel K, Wang J (2005) Interaction of organic cations with a newly identified plasma membrane monoamine transporter. Mol Pharmacol 68:1397–1407
113. Dahlin A, Xia L, Kong W, Hevner R, Wang J (2007) Expression and immunolocalization of the plasma membrane monoamine transporter in the brain. Neuroscience 146:1193–1211
114. Gasser PJ, Orchinik M, Raju I, Lowry CA (2009) Distribution of organic cation transporter 3, a corticosterone-sensitive monoamine transporter, in the rat brain. J Comp Neurol 512:529–555
115. Baganz NL, Horton RE, Calderon AS, Owens WA, Munn JL, Watts LT, Koldzic-Zivanovic N, Jeske NA, Koek W, Toney GM, Daws LC (2008) Organic cation transporter 3: keeping the break on extracellular serotonin in serotonin-transporter-deficient mice. Proc Nat Acad Sci USA 105:18976–18981
116. Feng N, Mo B, Johnson PL, Orchinik M, Lowry CA, Renner KJ (2005) Local inhibition of organic cation transporters increases extracellular serotonin in the medial hypothalamus. Brain Res 1063:69–76

117. Vialou V, Balasse L, Callebert J, Launay J-M, Giros B, Gautron S (2008) Altered aminergic neurotransmission in the brain of organic cation transporter 3-deficient mice. J Neurochem 106:1471–1482
118. Wultsch T, Grimberg G, Schmitt A, Painsipp E, Wetzstein H, Breitenkamp AFS, Gründemann D, Schömig E, Lesch K-P, Gerlach M, Reif A (2009) Decreased anxiety in mice lacking the organic cation transporter 3. J Neural Transm 116:689–697
119. Westkaemper RB, Roth BL (2006) Structure and function reveal insights in the pharmacology of 5-HT receptor subtypes. In: Roth BL (ed) The serotonin receptors: from molecular pharmacology to human therapeutics. Human Press, Totowa, pp 39–58
120. Davies MA, C-y Chang, Roth BL (2006) Polymorphic and posttranscriptional modifications of 5-HT receptor structure: functional and pathological implications. In: Roth BL (ed) The serotonin receptors: from molecular pharmacology to human therapeutics. Humana Press, Totowa, pp 59–90
121. Barnes NM, Sharp T (1999) A review of central 5-HT receptors and their function. Neuropharmacology 38:1083–1152
122. Raymond JR, Turner JH, Gelasco AK, Ayiku HB, Coaxum SD, Arthur JM, Garnovskaya MN (2006) 5-HT receptor signal transduction pathways. In: Roth BL (ed) The serotonin receptors: from molecular pharmacology to human therapeutics. Humana Press, Totowa, pp 143–206
123. Bockaert J, Claeysen S, Dumuis A, Marin P (2010) Classification and signaling characteristics of 5-HT receptors. In: Müller C, Jacobs B (eds) Handbook of behavioral neurobiology of serotonin. Elsevier B. V, Amsterdam, pp 103–121
124. Guimarães FS, Carobrez AP, Graeff FG (2008) Modulation of anxiety behaviors by 5-HT-interacting drugs. In: Blanchard RJ, Blanchard DC, Griebel G, Nutt D (eds) Handbook of anxiety and fear. Elsevier B. V, Amsterdam, pp 241–268
125. Bechtholt AJ, Lucki I (2006) Effects of serotonin-related gene deletion on measures of anxiety, depression, and neurotransmission. In: Roth BL (ed) The serotonin receptors: from molecular pharmacology to human therapeutics. Humana Press, Totowa, pp 577–606
126. Riad M, Garcia S, Watkins KC, Jodoin N, Doucet É, Langlois X, El Mestikawy S, Hamon M, Descarries L (2000) Somatodendritic localization of 5-HT1A and preterminal axonal localization of 5-HT1B serotonin receptors in adult rat brain. J Comp Neurol 417:181–194
127. McDevitt RA, Neumaier JF (2011) Regulation of dorsal raphe nucleus function by serotonin autoreceptors: a behavioral perspective. J Chem Neuroanat 41:234–246
128. Mengod G, Vilaró MT, Cortés R, López-Giménez JF, Raurich A, Palacios JM (2006) Chemical neuroanatomy of 5-HT receptor subtypes in the mammalian brain. In: Roth BL (ed) The serotonin receptors: from molecular pharmacology to human therapeutics. Humana Press, Totowa, pp 319–364
129. Mengod G, Cortés R, Vilaró MT, Hoyer D (2010) Distribution of 5-HT receptors in the central nervous system. In: Müller C, Jacobs B (eds) Handbook of behavioral neurobiology of serotonin. Elsevier B. V, Amsterdam, pp 123–138
130. Clarke WP, Yocca FD, Maayani S (1996) Lack of 5-hydroxytryptamine$_{1A}$-mediated inhibition of adenylyl cyclase in dorsal raphe of male and female rats. J Pharmacol Exp Ther 277:1259–1266
131. Ramboz S, Oosting R, Amara DA, Kung HF, Blier P, Mendelsohn M, Mann JJ, Brunner D, Hen R (1998) Serotonin receptor 1A knockout: an animal model of anxiety-related disorder. Proc Nat Acad Sci USA 95:14476–14481
132. Gross C, Zhuang X, Stark K, Ramboz S, Oosting R, Kirby L, Santarelli L, Beck S, Hen R (2002) Serotonin$_{1A}$ receptor acts during development to establish normal anxiety-like behaviour in the adult. Nature 416:396–400
133. Richardson-Jones JW, Craige CP, Guiard BP, Stephen A, Metzger KL, Kung HF, Gardier AM, Dranovsky A, David DJ, Beck SG, Hen R, Leonardo ED (2009) 5-HT$_{1A}$ autoreceptor levels determine vulnerability to stress and response to antidepressants. Neuron 65:40–52
134. Richardson-Jones JW, Craige CP, Nguyen TH, Kung HF, Gardier AM, Dranovsky A, David DJ, Guiard BP, Beck SG, Hen R, Leonardo ED (2011) Serotonin-1A autoreceptors are

necessary and sufficient for the normal formation of circuits underlying innate anxiety. J Neurosci 31:6008–6018
135. Albert PR, Morris SJ, Ghahremani MH, Storring JM, Lembo PMC (1998) A putative α-helical G $\beta\gamma$-coupling domain in the second intracellular loop of the 5-HT_{1A} receptor. Ann N Y Acad Sci 861:146–161
136. Zhong H, Neubig RR (2001) Regulator of G protein signaling proteins: novel multifunctional drug targets. J Pharmacol Exp Ther 297:837–845
137. Talbot JN, Jutkiewicz EM, Graves SM, Clemans CF, Nicol MR, Mortensen Rd, Huang X-Y, Neubig RR, Traynor JR (2010) RGS inhibition at $G\alpha_{i2}$ selectively potentiates 5-HT1A-mediated antidepressant effects. Proc Nat Acad Sci USA 107:11086–11091
138. Ou XM, Storring JM, Kushwaha N, Albert PR (2001) Heterodimerization of mineralocorticoid and glucocorticoid receptors at a novel negative response element of the 5-HT_{1A} receptor gene. J Biol Chem 276:14299–14307
139. Meijer OC, Williamson A, Dallman MF, Pearce D (2000) Transcriptional repression of the 5-HT_{1A} receptor promoter by corticosterone via mineralocorticoid receptors depends on the cellular context. J Neuroendocrinol 12:245–254
140. Mendelson SD, McEwen BS (1992) Autoradiographic analyses of the effects of adrenalectomy and corticosterone on 5-HT_{1A} and 5-HT_{1B} receptors in the dorsal hippocampus and cortex of the rat. Neuroendocrinology 55:444–450
141. Chalmers DT, Kwak SP, Mansour A, Akil H, Watson SJ (1993) Corticosteroids regulate brain hippocampal 5-HT_{1A} receptor mRNA expression. J Neurosci 13:914–923
142. Zhong P, Cianarello RD (1995) Transcriptional regulation of hippocampal 5-HT_{1A} receptors by corticosteroid hormones. Mol Brain Res 29:23–34
143. Tejani-Butt SM, Labow DM (1994) Time course of the effects of adrenalectomy and corticosterone replacement on 5-HT_{1A} and 5-HT uptake sites in the hippocampus and dorsal raphe nucleus of the rat brain: an autoradiographic analysis. Psychopharmacology 113:481–486
144. Laaris N, Le Poul E, Laporte AM, Hamon M, Lanfumey L (1999) Differential effects of stress on presynaptic and postsynaptic 5-hydroxytryptamine-1A receptors in the rat brain: an in vitro electrophysiological study. Neuroscience 91:947–958
145. Liao B, Miesak B, Azmitia EC (1993) Loss of 5-HT_{1A} receptor mRNA in the dentate gyrus of the long-term adrenalectomized rats and rapid reversal by dexamethasone. Mol Brain Res 19:328–332
146. Meijer OC, de Kloet ER (1994) Corticosterone suppresses the expression of 5-HT_{1A} receptor mRNA in rat dentate gyrus. Eur J Pharmacol 266:255–261
147. Meijer OC, de Kloet ER (1995) A role for the mineralocorticoid receptor in a rapid and transient suppression of hippocampal 5-HT_{1A} receptor mRNA by corticosterone. J Neuroendocrinol 7:653–657
148. Wissink S, Meijer O, Pearce D, van der Burg B, van der Saag PT (2000) Regulation of the rat serotonin-1A receptor gene by corticosteroids. J Biol Chem 275:1321–1326
149. Laaris N, Haj-Dahmane S, Hamon M, Lanfumey L (1995) Glucocorticoid receptor-mediated inhibition by corticosterone of 5-HT_{1A} autoreceptor functioning in the rat dorsal raphe nucleus. Neuropharmacology 34:1201–1210
150. Laaris N, Le Poul E, Hamon M, Lanfumey L (1997) Stress-induced alterations of somatodendritic 5-HT_{1A} autoreceptor sensitivity in the rat dorsal raphe nucleus—in vitro electrophysiological evidence. Fundam Clin Pharmacol 11:206–214
151. Lanfumey L, Pardon MC, Laaris N, Joubert C, Hanoun N, Hamon M, Cohen-Salmon C (1999) 5-HT_{1A} autoreceptor desensitization by chronic ultramild stress in mice. NeurorReport 10:3369–3374
152. Fairchild G, Leitch MM, Ingram CD (2003) Acute and chronic effects of corticosterone on 5-HT_{1A} receptor-mediated autoinhibition in the rat dorsal raphe nucleus. Neuropharmacology 45:925–934

153. Judge SJ, Ingram CD, Gartside SE (2004) Moderate differences in circulating corticosterone alter receptor-mediated regulation of 5-hydroxytryptamine neuronal activity. J Psychopharmacol 18:475–483
154. Lanfumey L, Mongeau R, Cohen-Salmon C, Hamon M (2008) Corticosteroid-serotonin interactions in the neurobiological mechanisms of stress-related disorders. Neurosci Biobehav Rev 32:1174–1184
155. Ou XM, Lemonde S, Jafard-Nejad H, Bown CD, Goto A, Rogaeva A, Albert PR (2003) Freud-1: a neuronal calcium-regulated repressor of the 5-HT1A receptor gene. J Neurosci 23:7415–7425
156. Hadjighassem MR, Austin MC, Szewczyk B, Daigle M, Stockmeier CA, Albert PR (2009) Human Freud-2/CC2D1B: a novel repressor of postsynaptic serotonin-1A receptor expression. Biol Psychiatry 66:214–222
157. Iyo AH, Kieran N, Chandran A, Albert PR, Wicks I, Bissette G, Austin MC (2009) Differential regulation of the 5-HT_{1A} transcriptional modulators Freud-1 and NUDR by chronic stress. Neuroscience 163:1119–1127
158. Kieran N, Ou XM, Iyo AH (2010) Chronic social defeat downregulates the 5-HT1A receptor but not Freud-1 or NUDR in the rat prefrontal cortex. Neurosci Lett 469:380
159. Ramboz S, Saudou F, Amara DA, Belzung C, Segu L, Misslin R, Buhot MC, Hen R (1996) 5-HT1B receptor knock out—behavioral consequences. Behav Brain Res 73:305–312
160. Ase AR, Reader TA, Hen R, Riad M, Descarries L (2000) Altered serotonin and dopamine metabolism in the CNS of serotonin 5-HT_{1A} or 5-HT_{1B} receptor knockout mice. J Neurochem 75:2415–2426
161. Clark MS, Sexton TJ, McClain M, Root DC, Kohen R, Neumaier JF (2002) Overexpression of 5-HT1B receptor in dorsal raphe nucleus using herpes simplex virus gene transfer increases anxiety behavior after inescapable stress. J Neurosci 22:4550–4562
162. Clark MS, Vincow ES, Sexton TJ, Neumaier JF (2004) Increased expression of 5-HT_{1B} receptor in dorsal raphe nucleus decreases fear-potentiated startle in a stress dependent manner. Brain Res 1007:86–97
163. McDevitt RA, Hiroi R, Mackenzie SM, Robin NC, Cohn A, Kim JJ, Neumaier JF (2011) Serotonin 1B autoreceptors originating in the caudal dorsal raphe nucleus reduce expression of fear and depression-like behavior. Biol Psychiatry 69:780–787
164. Neumaier JF, Edwards E, Plotsky PM (2002) 5-HT_{1B} mRNA regulation in two animal models of altered stress reactivity. Biol Psychiatry 51:902–908
165. Svenningsson P, Chergui K, Rachleff I, Flajolet M, Zhang X, El Jacoubi M, Vaugeois J-M, Nomikos GG, Greengard P (2006) Alterations in 5-HT_{1B} receptor function by p11 in depression-like states. Science 311:77–80
166. Serrats J, Mengod G, Cortes R (2005) Expression of serotonin 5-HT2C receptors in GABAergic cells of the anterior raphe nuclei. J Chem Neuroanat 29:83–91
167. Heisler LK, Zhou L, Bajwa P, Hsu J, Tecott LH (2007) Serotonin 5-HT_{2C} receptors regulate anxiety-like behavior. Genes Brain Behav 6:491–496
168. Pigott TA, Hill JL, Grady TA, L'Hereux F, Bernstein S, Rubenstein CS, Murphy DL (1993) A comparison of the behavioral effects of oral versus intravenous *m*CPP administration in OCD patients and the effect of metergoline prior to IV *m*CPP. Biol Psychiatry 33:3–14
169. Klein E, Zohar J, Gerazi MF, Murphy DL, Uhde TW (1991) Anxiogenic effects of *m*-CPP in patients with panic disorder: comparison to caffeine's anxiogenic effects. Biol Psychiatry 30:973–984
170. Germine M, Goddard AW, Woods SW, Charney DS, Heninger GR (1992) Anger and anxiety responses to *m*-chlorophenylpiperazine in generalized anxiety disorder. Biol Psychiatry 32:457–461
171. Griebel G (1995) 5-hydroxytryptamine-interacting drugs in animal models of anxiety disorders: more than 30 years of research. Pharmacol Ther 65:319–395
172. Gatch MB (2003) Discriminative stimulus effects of m-chlorophenylpiperazine as a model of the role of serotonin receptors in anxiety. Life Sci 73:1347–1367

173. Berg KA, Maayani S, Goldfarb J, Scaramellini C, Leff P, Clarke WP (1998) Effector pathway-dependent relative efficacy at serotonin type 2A and 2C receptors: evidence for agonist-directed trafficking of receptor stimulus. Mol Pharmacol 54:94–104
174. Burns CM, Chu H, Rueter SM, Hutchinson LK, Canton H, Sanders-Bush E, Emeson RB (1997) Regulation of serotonin-2C receptor G-protein coupling by RNA editing. Nature 387:303–308
175. Niswender CM, Copeland SC, Herrick-Davis K, Emeson RB, Sanders-Bush E (1999) RNA editing of the human serotonin 5-hydroxytryptamine 2C receptor silences constitutive activity. J Biol Chem 274:9472–9478
176. Herrick-Davis K, Grinde E, Niswender CM (1999) Serotonin 5-HT2C receptor RNA editing alters receptor basal activity: implications for serotonergic signal transduction. J Neurochem 73:1711–1717
177. Marion S, Weiner DM, Caron MG (2004) RNA editing induces variation in desensitization and trafficking of 5-hydroxytryptamine 2c receptor isoforms. J Biol Chem 279:2945–2954
178. Quirk K, Lawrence A, Jones J, Misra A, Harvey V, Lamb H (2001) Characterisation of agonist binding on human 5-HT2C receptor isoforms. Eur J Pharmacol 419:107–112
179. Wang Q, O'Brien PJ, Chen CX, Cho DS, Murray JM, Nishikura K (2000) Altered G protein-coupling functions of RNA editing isoform and splicing variant serotonin2C receptors. J Neurochem 74:1290–1300
180. Berg KA, Cropper JD, Niswender CM, Sanders-Bush E, Emeson RB, Clarke WP (2001) RNA-editing of the 5-HT_{2C} receptor alters agonist-receptor-effector coupling specificity. Br J Pharmacol 134:386–392
181. Dracheva S, Lyddon R, Barley K, Marcus SM, Hurd YL, Byne WM (2009) Editing of serotonin 2C receptor mRNA in the prefrontal cortex characterizes high-novelty locomotor response behavioral trait. Neuropsychopharmacology 34:2237–2251
182. Hackler EA, Airey DC, Shannon CC, Sodhi MS, Sanders-Bush E (2006) 5-HT_{2C} receptor RNA editing in the amygdala of C57BL/6J, DBA/2J, and BALB/cJ mice. Neurosci Res 55:96–104

Chapter 3
Nodal Structures in Anxiety-Like and Panic-Like Responses

3.1 Nodal Structures Regulating Anxiety: The Behavioral Inhibition System

Two structures have been proposed as central in the control of anxiety-like responses, the basolateral amygdala [1] and the ventral hippocampus [2]; in addition, great attention is being given recently to the role of the lateral habenula [3–5] and cingulate cortex [6–8] in this process. The cingulate cortex, ventral hippocampus (VH), and lateral habenula (LHb)—Structures which are located upstream in this *behavioral inhibition system* [9, 10]—Sample multimodal sensory information in order to detect mismatches between expected and actual outcomes [9, 11]. When a mismatch is detected (i.e., in unpredictable situations [12]), these regions send input to the basolateral amygdala (BLA), which assesses the emotional salience of stimuli, biasing behavior toward avoidance [9, 13] or inhibiting affectively positive memories [14]. Projections from the BLA to the central amygdala (CeA) and bed nucleus of the stria terminalis (BNST) regulate autonomic and behavioral responses to uncertain aversive stimuli [1, 15, 16]. The lateral habenula also occupies an important neuroanatomical position that could warrant a role in this "behavioral inhibition system", since projections from the LHb inhibits dopaminergic neurons in the ventral tegmental area and in the substantia nigra pars compacta and stimulates serotonergic neurons from the raphe [11].

3.2 "Limbic" Portions of the Medial Prefrontal Cortex

The medial prefrontal cortex, especially its ventral portions, has been implicated in fear- and anxiety-like responses in diverse paradigms. We follow the definition proposed by Vogt and colleagues [17] that a "limbic" cortex is every cortical structure which has specific roles in the regulation of autonomic responses, dense projections to the hypothalamus, and which subserves emotions. These include the

C. Maximino, *Serotonin and Anxiety*, SpringerBriefs in Neuroscience,
DOI: 10.1007/978-1-4614-4048-2_3, © The Author(s) 2012

perigenual anterior and caudal cingulate cortices (pACC and MCC), retrosplenial cortex, and infralimbic and prelimbic cortices. These regions receive massive diffuse projections from the anteromedial thalamic nucleus [17, 18], a subdivision of the thalamus which mediates the reciprocal connections between mPFC and hippocampus [19], and the pACC receives an important projection from the basolateral amygdala [20]. A direct projection from the ventral hippocampus is found exclusively in the prelimbic cortex [21]. Projections from infralimbic and prelimbic cortices differ; IL projection sites are the lateral septum, bed nucleus of the stria terminalis, medial and lateral preoptic nuclei, substantia inominata, intercalated masses and medial and central amygdala, mesopontine rostromedial tegmental nucleus, and dorsomedial and lateral hypothalamus, while the PL projects to the insular cortex, claustrum, nucleus accumbens, central amygdala "shell", mesopontine rostromedial tegmental nucleus, and dorsal raphe [22–28].

Limbic cortical regions have been implicated in the control of anxiety-like responses in studies ranging from human neuroimaging to animal lesion studies. In patients with generalized anxiety disorder, the anterior cingulate cortex is more activated during the presentation of cues which predict aversive pictures than in healthy controls, but this "anticipatory" activation is also greater during the presentation of cues which predict neutral pictures; interestingly, greater ACC activation predicts greater reductions in anxiety and worry symptoms after eight weeks of venlafaxine treatment [29]. In another study, healthy voluntaries were trained to associate a cue with small electrical shocks of four intensities; ACC activation was weakly correlated with self-reported anxiety level at low-to-moderate shock intensities, and negatively and strongly correlated with anxiety level at strong shock intensities [30]. Yet another study found that such procedures *reduce* blood flow in the ACC, an effect which is inversely and moderately correlated with anxiety self-rating [31]. In rats selectively bred for high anxiety-like behavior in the elevated plus-maze (HABs), similar effects are observed in that attenuation of c-Fos-like immunoreactivity in the cingulate cortex follows exposure to an open-field, to the open arms of an elevated plus-maze, after social defeat, and after administration of FG-7142; no effect is observed after airjet stress, a manipulation which evokes escape reactions [32–34]. Likewise, animals which have been previously exposed to footshock in the home cage shift fleeing to freezing when exposed to an anxiogenic ultrasound ($\sim$20 kHz); these animals also show decreased c-Fos-like immunoreactivity in the retrosplenial, anterior cingulate, and prelimbic cortices [35]. These results suggest that the degree of activity in the ACC is inversely related to anxiety level. Surprisingly, in rats exposed to novelty stress, the availability of chewing material (which leads to "displacement" chewing behavior, a strategy of coping) in a novel environment increases c-Fos-like immunoreactivity in the right (but not left) mPFC while simultaneously decreasing activity in the right (but not left) central amygdala [36], a phenomenon which is associated with decreased dopamine turnover in the right mPFC [37].

Experiments analyzing the effect of lesions have been performed mainly on rats. Ibotenic acid lesions of the right, but not left, infralimbic and prelimbic cortices are anxiolytic in the EPM, and right-lesioned rats drink more of a sweetened milk/quinine solution despite not differing in consumption of sweetened milk alone [7]. Bilateral ibotenic acid microinjections in the infralimbic and prelimbic cortices is also anxiolytic in the open field, increases freezing to contextual cues and reduces freezing to discrete cues [38]. Bilateral electrolytic lesions of the cingulate cortex decreases inhibitory avoidance in the elevated T-maze, increasing escape latencies only after 3 h of restraint stress [39]. Lesions of the prelimbic cortex enhance restraint-induced c-Fos-like immunoreactivity and CRF mRNA expression in autonomic cells of the dorsomedial parvocellular portion of the paraventricular hypothalamus. Lesions in the infralimbic cortex, on the other hand, decreases restraint-induced c-Fos-like immunoreactivity and CRF mRNA in this compartment, while at the same time increasing c-Fos-like immunoreactivity in the ventromedial parvocellular portion of the paraventricular hypothalamus [6].

An important role for the detection of stressful uncontrollable stimuli has been proposed. mPFC neurons project heavily to the dorsal raphe nucleus (especially the dorsal portion of the DRN), where mPFC activation leads to raphe inhibition. This inhibition is usually followed by rebound increases in the probability of DRN cell firing, suggesting that cortical stimulation entrains the spiking of serotonergic cells in the DRN, effectively synchronizing the activity of these cells [40]. These projections have been proposed to regulate the responsiveness of serotonergic neurons to controllable versus uncontrollable stress. Muscimol microinjections in the PL and IL during uncontrollable and inescapable shocks do not have effects on behavior or 5-HT activation. When injections are made in a schedule of controllable and escapable shocks, on the other hand, DRN activation is increased up to the level produced by uncontrollable stress, and an increase in anxiety-like behavior and impairment of escape learning (i.e., effects observed after uncontrollable stress) is observed [41]. If animals are exposed to controllable stress before being exposed to uncontrollable shocks, the effects of uncontrollable stress are blocked, an "immunization" effect which is prevented by muscimol micro-injections before or during immunization [42]. Microinjections of picrotoxin during uncontrollable stress eliminate behavioral effects and DRN activation [43]. Finally, controllable stress activates PL neurons which project to the DRN [44].

mPFC neurons and astrocytes also receive important projections from the dorsal raphe, especially from the subregion known as the interfascicular part (DRI), a region that is responsive to systemic bacterial infections by increasing 5-HT release in the prelimbic and infralimbic cortices [45, 46]. Serotonin release in the medial prefrontal cortex is increased in mice exposed to predators in an inescapable situation [47], and rats which have been selected for low sensitivity to acute painful stimulation show active responses to a fear-conditioned context that are concordant with stronger serotonin immunoreactivity in the medial prefrontal cortex than passive-coping, highly sensitive animals [48]. Likewise, exposure of guinea pigs to an EPM increases serotonin release in the prefrontal cortex, an effect which is potentiated by cholecystokinin receptor B (CCK-B) agonists [49–51]; these drugs also produce a marked anxiogenic profile. CCK-B agonists do not affect anxiety-like behavior or 5-HT levels when animals are observed in the home cage [49], and their effect is blocked by systemic administration of an anxiolytic dose of 8-OH-DPAT

[50] or diazepam [52], suggesting that CCK-B receptors mediate changes in 5-HT release under aversive conditions, but not in resting states.

In the mPFC, serotonin regulates the synchronization of neuronal ensemble activity that is controlled by fast-spiking interneurons. The majority of fast-spiking cells are inhibited by serotonin via 5-HT_{1A}Rs, while a small amount is activated by serotonin via 5-HT_{2A}Rs, and 5-HT_{2A}Rs regulate the frequency and amplitude of slow and fast rhythms in the mPFC [53]. It is unknown if and how these effects mediate the control of anxiety by the mPFC, but serotonin depletion in this region (by local microinjection of the serotonergic neurotoxin 5,7-dihydroxytryptamine) increases anxiety-like behavior in the elevated plus-maze [54]. Rats which show divergent reactivity to novel environments also show divergent proportions of an edited isoform of the 5-HT_{2C} receptor, with animals presenting lower levels of locomotor activity in a novel situation also presenting lower levels of post-transcriptionally edited receptors in the prefrontal cortex [55]. Likewise, BALB/c mice, a strain which shows lower forebrain serotonin levels [56], increased anxiety, and increased stress reactivity in relation to C57BL/6 and 129Sv mice [57, 58]; the majority of cortical 5-HT_{2C} mRNA is nonedited [59]. In this strain, acute swim stress elicits pre-mRNA editing in 5-HT_{2C} receptors in the medial prefrontal cortex, an effect which is blocked by chronic fluoxetine treatment; these latter effects are not observed in C57BL/6 mice, in which the majority of cortical 5-HT_{2C} mRNA is edited [59].

3.3 The Extended Amygdala

For a long time, the basolateral amygdala (BLA) has been associated with anxiety-like behavior and fear [60–64]. The BLA receives important projections from the piriform, entorhinal, orbital, prelimbic, insular, and cingulate cortices, as well as from the nucleus accumbens and CA1 layer of the hippocampus [65]. Important projections also derive from a diverse array of thalamic nuclei, as well as from the "olfactory" aspects of the amygdala (i.e., the medial extended amygdala) and from the lateral amygdaloid nucleus [65], strongly supporting a sensorial interfacing role for the BLA [63, 66]. "Neuromodulatory" input comes from dopaminergic ventral tegmental area neurons, serotonergic caudal linear nucleus, nucleus raphe magnus, and dorsomedial raphe nucleus shell neurons [65]. Widespread glutamatergic projections to the dorsal and ventral striatum arise from the BLA [67], with a denser distribution of terminals in the ventral striatum, an intermediate amount in the dorsomedial parts of the striatum, and no terminals in the dorsolateral portions. BLA neurons also receive one of the densest cholinergic projections of all brain nuclei, stemming from the basal forebrain [65]. The main targets of the BLA are the central amygdala (CeA) and the bed nucleus of the stria terminalis (BNST), with neurons projecting to each of these portions not overlapping in the BLA [15].

Sensitization of BLA circuitry by repeated epileptogenic stimulation increases risk assessment in the elevated plus-maze without changing open-arm avoidance

[68]. In kindled rats, expression of the astrocytic glutamate/aspartate transporter (GLAST) is decreased in the amygdala [69]; consistently, the chronic blockade of astrocytic glutamate transporters (thereby promoting a chronic increase in extrasynaptic glutamate concentrations) in the BLA decreases social exploration in rats [70]. Exposure to an open field under low-light conditions increases c-Fos-like immunoreactivity in the BLA of rats, while high-light conditions selectively increase c-Fos-like immunoreactivity in the posterior and ventromedial portions of the BLA [71], as well as in ventral hippocampal [72] and dorsomedial and caudodorsal raphe cells [73] which project to the basolateral amygdala. The administration of anxiogenic drugs with multiple mechanisms (the 5-HT_2R non-selective agonist *m*-chlorophenyl piperazine [mCPP], the adenosine non-selective antagonist caffeine, the α_2-adrenoceptor antagonist yohimbine, and the partial benzodiazepine receptor inverse agonist *N*-methyl-beta-carboline-3-carboxamide [FG-7142]) increase c-Fos-like immunoreactivity in parvalbumin-positive GABAergic neurons of the BLA [74]. Activation of the BLA has also been observed in mice receiving aversive ultrasound in the home cage, and this Fos expression is not changed when footshocks are administered before ultrasound exposure [35]. Moreover, the BLA is activated after electrical stimulation of the inferior colliculus by currents which elicit freezing as well as by currents which elicit escape responses [75]. Since these manipulations shift the balance from flight (active fear responses) to freezing (passive fear responses), or from high locomotion to low locomotion, this suggests that the BLA has no role in the selection of behavioral strategies; instead, the switch is found downstream in the cerebral aversion system.

In rats conditioned to freeze to a context, the extent of freezing is tightly correlated with increases in protein kinase C βIII in the right BLA [76], and greater serotonin content in the right versus left BLA is directly correlated with anxiety-like behavior in the plus-maze [77]. Muscimol microinjections in the right BLA block retention of inhibitory avoidance in which freezing is a major component [78, 79]. Likewise, in animals which were exposed to a predator, blockade of NMDA receptors in the right BLA impairs decreases in startle latency, while blockade of NMDA receptors in the left BLA impairs increases in risk assessment in the elevated plus-maze and startle amplitude [80]. Exposure to predators also produces long lasting increases of potentials recorded in the BLA after hippocampal stimulation in both hemispheres of the brain, an effect which is blocked in the right hemisphere and amplified in the left hemisphere by systemic administration of NMDA antagonists [81]. Thus, diminishing the activity of the right BLA, leaving the *left* BLA active, contributes to a reduction in passive responses (freezing) or to behavioral activation, and thus BLA laterality could be an important substrate for the passive-versus-active coping distinction. A role for BLA plasticity in establishing long lasting alterations in defensive behavior after predator stress has also been proposed. Repeated evocation of seizures in the right basolateral amygdala (kindling) produces anxiogenic effects in animals with low baseline anxiety, and anxiolytic effects in animals with high baseline anxiety; left BLA kindling is either anxiogenic in low baseline anxiety animals, or ineffective in high baseline anxiety animals [82]. In animals which show increased anxiety after left BLA kindling, 5-HT$_{1A}$ receptor binding and mRNA expression in the dentate gyrus is increased [83]. The depression observed in this left VH-BLA pathway is more short-lived than potentiation in the right VH-BLA pathway [80], suggesting that it has an initiatory, but not sustaining, role in lasting behavioral inhibition following anxiogenic manipulations.

Serotonergic innervation of the basolateral amygdala originates mainly in the mid-rostrocaudal region of the dorsal part of the dorsal raphe nucleus [73, 84, 85]. Exposure of rats to an open field elicits c-Fos-like expression in diverse neurons of the BLA [71, 72], including neurons which receive projections from the ventral hippocampus [72] and from the DRN [73], and inescapable stress and microinjection of the CRF$_2$R agonist urocortin 2 in the DRN increase serotonin release in the BLA [86, 87]. Anxiogenic drugs with different molecular targets (mCPP, caffeine, yohimbine, FG-7142) activate GABAergic interneurons in the BLA which expresses parvalbumin [74]. These interneurons express high levels of the 5-HT$_{2A}$ receptor [88, 89]; the increased serotonergic efflux in the BLA, thus, leads to increased GABA release by these interneurons [89, 90]. GABA then acts to inhibit glutamate-evoked action potentials in the principal neurons of the BLA [91]. This effect seems to be dependent on circulating corticosteroids, as adrenalectomy abolishes serotonergic inhibition of glutamate-evoked action potentials in the BLA [91]. Concurrently, 5-HT$_{1A}$ receptors in principal neurons depress excitatory transmission and calcium responses [92], and 8-OH-DPAT microinjections in this region produce anxiolytic (but not panicolytic) effects in the elevated T-maze [93]. The overall result is a marked decrease in BLA output, decreasing anxiety-like responses (Figs. 3.1 and 3.2). Nonetheless, the microinjection of serotonin in the BLA facilitates inhibitory avoidance in the elevated T-maze, without effects on escape responses [94], and the destruction of serotonergic fibers in the BLA with the neurotoxin 5,7-dihydroxytryptamine (5,7-DHT) releases punished behavior in the Vogel test, but spares anxiety-like behavior in the elevated plus-maze [95]. Since, the activation of 5-HT$_{2C}$ receptors in the BLA also produces an anxiogenic effect [94, 96–99], and the blockade of these receptors is anxiolytic (but not panicolytic) in the elevated T-maze [94], it is possible that the serotonergic tone on BLA 5-HT$_{2C}$Rs is predominant on the control of anxiety-like behavior by this structure. Moreover, DBA/2J and BALB/cJ mice show a reduction in *htr2c* pre-mRNA editing in the amygdala in relation to C57BL/6J mice that is concordant with increased anxiety in the first in relation to the latter [100], suggesting that 5-HT$_{2C}$ receptor trafficking and/or effector coupling is important in the regulation of anxiety-like behavior by the BLA.

3.4 The Ventral Hippocampus

Convergent evidence for a specialization of the ventral hippocampus for the control of anxiety has appeared in the last decade [2]. The ventral and dorsal hippocampi show different patterns of gene expression [101, 102], with gene

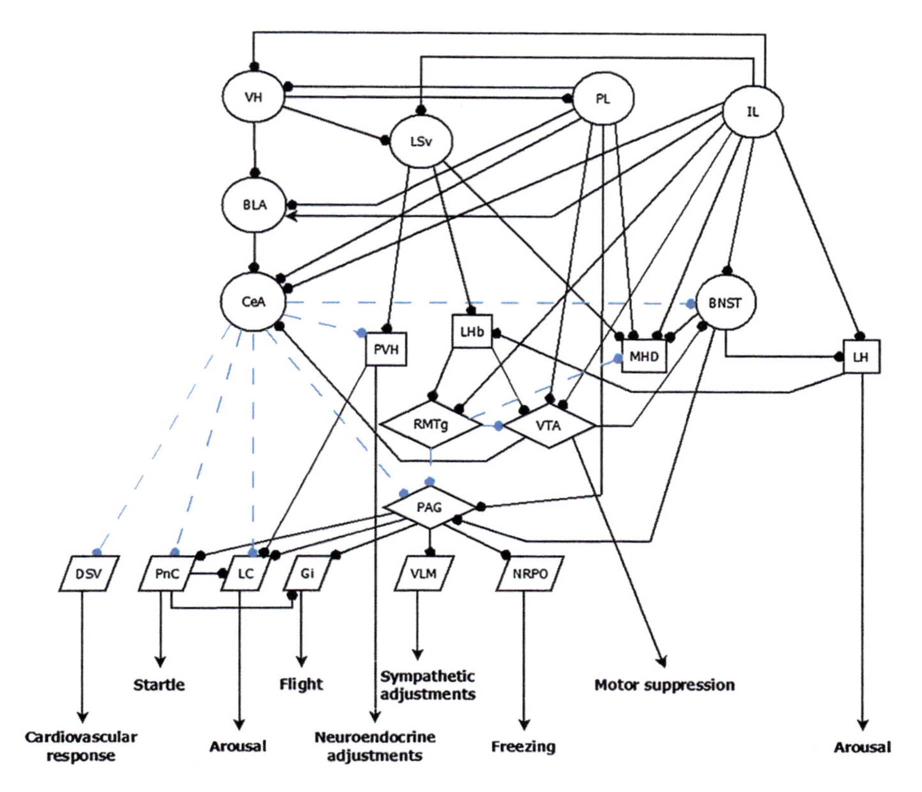

Fig. 3.1 Nodal structures controlling fear and anxiety responses. Telencephalic structures are represented by *circles*, diencephalic structures are represented by *rectangles*, mesencephalic structures are represented by *losanges*, and rhombencephalic structures are represented by *paralelograms*. Also marked are the different behavioral outputs associated with fear- and anxiety-like states. Excitatory projections are marked by full *black lines*, while inhibitory connections are marked by *dashed blue lines*. Abbreviations: *BLA* basolateral amygdala, *BNST*, bed nucleus of the stria terminalis, *CeA* central amygdala, *DSV* dorsal vagal complex,*Gi* nucleus reticularis gigantocelularis, *IL* infralimbic cortex, *LC* locus coeruleus, *LH* lateral hypothalamus, *LHb* lateral habenula, *LSv* ventral part of the lateral septum, *MHD* medial hypothalamic defensive system, *NRPO* nucleus reticularis pontis oralis, *PAG* periaqueductal gray, *PL* prelimbic cortex, *PnC* nucleus reticularis pontis caudalis, *PVH* paraventricular hypothalamus, *RMTg* rostromedial tegmental nucleus, *VH* ventral hippocampus, *VLM* ventrolateral medulla

expression in the ventral hippocampus correlating with that in other areas specialized for defense reactions [103]. Acute stress leads to increases in LTP in the VH and suppression in the DH [104], and increases long-term depression (LTD) in DH while converting LTD to slow the onset of LTP in the VH [105]. Similarly, theta oscillations in the ventral hippocampus differentiate between open and closed arms of a radial maze, while theta rhythms in the dorsal hippocampus represent discrete place fields [106]. Knockout mice for the modifier subunit of glutamate cysteine ligase, the rate-limiting enzyme in the synthesis of the

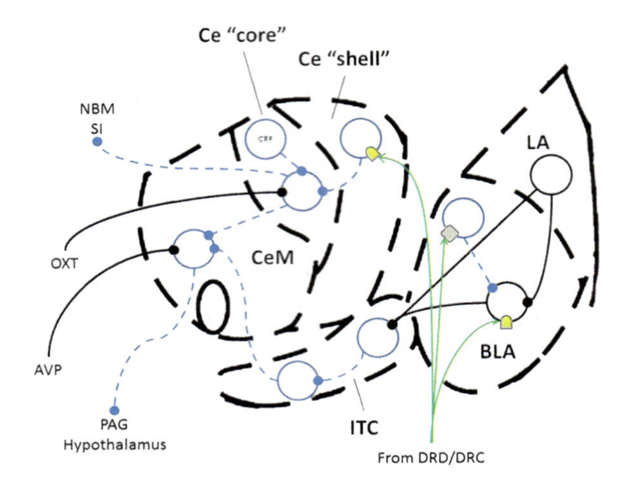

Fig. 3.2 Circuitry for the control of defensive responses in the amygdala. Neurons from the basolateral (BLA) and lateral amygdala (*LA*) do not project directly to the output cells of the medial central amygdala (CeM); instead, they send excitatory projections to interneurons of the intercalated islands, which inhibit output neurons from this cluster, resulting in disinhibition of CeM output cells. 5-HT_{1A}Rs (*yellow shapes*) in glutamatergic neurons of the BLA decrease their activity, resulting in inhibition of CeM output cells; in parallel, 5-HT_{2A}Rs (*gray rectangles*) in GABAergic interneurons of the BLA increase the inhibitory tone on glutamatergic cells, further increasing the inhibition of CeM output neurons. Type I GABAergic neurons in the CeA "shell" express 5-HT_{1A}Rs, which, when activated, turn the switch "off". In this situation, oxytocin receptor-expressing GABAergic Type II neurons in the CeA "core" are disinhibited, resulting in increased inhibition of CeM output neurons and releasing the tonically inhibited cholinergic cells of the nucleus basalis of Meynert and substantia innominata. The net result of turning the switch "off" is a shift from freezing to risk assessment. CRFergic projections to GABAergic Type II interneurons of the CeA "core" produce an opposite effect, disinhibiting CeM output neurons and increasing inhibition on cholinergic cells responsible for promoting cortical arousal. Based on Gozzi et al. [198] and Paré et al. [62]. Anatomic diagram from Paxinos and Watson [199]

endogenous antioxidant glutathione, causes a selective decrease of parvalbumin-immunoreactive neurons in the VH, but not in the DH, followed by a concomitant reduction of β/γ oscillations and decreased anxiety in the elevated plus-maze and light/dark preference test [107]. Lidocaine microinjections in the VH are anxiolytic in trial 1 in the elevated plus-maze, while microinjections in the DH are anxiolytic in trial 2 [108]. Lesions of the VH, but not of the DH, significantly impair conditioned and unconditioned defensive behavior in rats exposed to fear conditioning or predator stimuli [109]. Lesions of the VH produce anterograde deficits in conditioned freezing in a signaled shock procedure [110, 111] and impairs delay fear conditioning, an effect which is not observed in dorsal hippocampal lesions [112]. Ventral hippocampal theta activity correlates with medial prefrontal cortex theta, and this correlation increases in the open field and elevated plus-maze [8]. Moreover, VH theta activity differentiates between open and closed arms of a radial maze [106]. Exposure to an open field elicits c-Fos-like immunoreactivity in VH neurons which project to the basolateral amygdala [72]. Last, but not least, VH lesions show reduced hyponeophagia, increased social interaction, and decreased anxiety in the light/dark preference and successive alleys tests [113].

Serotonin in the ventral hippocampus originates in the caudal portion of the dorsal raphe nucleus [85], a region which contains neurons which are sensitive to anxiogenic drugs [114] and anxiogenic peptides [115–119]. Exposure to an elevated plus-maze [120, 121], predator stimuli [47] ,and inescapable shocks [86, 122] increases serotonin efflux in the VH, and rats exposed to previously inescapable shocks exhibit exaggerated 5-HT responses to two brief footshocks [86, 122]. Serotonin release in the hippocampus is mediated by adenosine A_1 receptors [123–125], which are under tonic activation by adenosine metabolized from astrocyte-released ATP [126]; this metabolic process is delayed after acute stress [127], which potentially reduces the inhibitory tonus on serotonin release.

The most studied 5-HT receptor in the hippocampus is 5-HT_{1A}R. This receptor produces hyperpolarization by a postsynaptic mechanism, followed by a depolarizing response mediated by 5-HT_4 receptor activation [128, 129]. The interplay between both activities mediates many complex functions of the VH. The synchronization between ventral hippocampal and medial prefrontal cortical theta rhythms in anxiogenic environments seems to be influenced by 5-HT_{1A} receptors, as knocking out this receptor increases the effect of anxiogenic environments on mPFC theta [8]. 5-HT_{1A}R knockout mice show enhanced fear conditioning to ambiguous conditioned stimuli [130]; conditional expression of 5-HT_{1A} receptors only in the dentate gyrus of the hippocampus rescues this phenotype [131], suggesting that 5-HT_{1A} receptors in the dentate gyrus regulate responses to ambiguous aversive stimuli. Microinjection of the 5-HT_{1A}R antagonist WAY 100635 into the VH is anxiolytic in the first trial of exposure to an elevated plus-maze, but not in the second trial [132]; however, VH-microinjected 5-HT_{1A}R agonists have no effect in the social interaction test or in the elevated plus-maze [133, 134], even though DH microinjection of these drugs release punished behavior [135]. These results are difficult to interpret, suggesting a serotonergic tone on ventral hippocampal 5-HT_{1A} receptors that produce anxiogenic effects.

Very few functional works analyzed the role of 5-HT_2 receptors in the ventral hippocampus. This represents a problem, since 5-HT_{2C} mRNA is enriched in the ventral portion of the CA1 domain of the hippocampus [103], and collateral efficacy has been observed in this region, with arachidonic acid release being elicited by 5-HT independent of phospholipase C activation [136]. Activation of the 5-HT_{2C} receptors in the VH, but not in the DH, is anxiogenic [137] (but see [138]). In 5-HT_{2C} receptor knockout mice, long-term potentiation in the perforant path-dentate gyrus synapses is impaired, a deficit that is accompanied by reduced latency to emerge from a dark chamber to a brightly lit one and reduced freezing to context [139]. Likewise, overexpression of 5-HT_{2C} receptors in forebrain regions (including hippocampus and amygdala, but not hypothalamus) produce a phenotype of increased anxiety-like behavior in mice [140]. These genetic experiments, however, are not anatomically specific, as these effects could be due to 5-HT_{2C} receptors in other forebrain structures.

The lateral septum is a non-homogeneous nucleus that shows topographically organized projections from the hippocampus, with its ventral pole receiving strong input from the VH [103]. The ventrolateral septum (LSv) is the mediator of a VH-medial hypothalamic projection that is very important in the control of neuroendocrine adjustments [141–143]. The microinjection of the benzodiazepine midazolam decreases avoidance latencies in the elevated T-maze, without effects on escape performance; contrariwise, the microinjection of the 5-HT_{1A} agonist 8-OH-DPAT into the LS caused an anxiogenic effect, increasing avoidance latencies, again without effect on escape latencies [144]. Similar results are observed in the elevated plus-maze and social interaction tests [145–147]. Thus, 5-HT_{1A} receptors in the lateral septum, a structure located downstream in the VH-hypothalamus circuit, show a clearer role in mediating the anxiogenic effects of serotonin in the septohippocampal system.

3.5 The Lateral Habenula

An emergent structure in the anxiety circuit is the habenula, which receives inputs from the septo-hippocampal system, the diagonal band of Broca, the lateral preoptic area, and lateral hypothalamus, and sends monosynaptic and disynaptic efferents to the serotonergic and dopaminergic systems [148]. The habenulae are important structures in the control of dopaminergic neurons associated with reward-seeking, and is also involved in fear, stress and anxiety responses [11]. The lateral portion of the habenula (LHb) is activated by stimuli which predict a small reward and inhibited by stimuli which predict a larger reward [149], a pattern that is opposed to what is observed in dopaminergic neurons [150]. Moreover, neurons in the LHb are also activated by stimuli which predict aversive consequences [151], and lesions in the LHb impair the acquisition of inhibitory avoidance in rats [152], and impair inhibitory avoidance acquisition and facilitate escape in the elevated T-maze [4]. Conversely, kainic acid microinjections in the LHb facilitate inhibitory avoidance and impair escape responses in the elevated T-maze [4, 5]; facilitation of inhibitory avoidance is blocked by concomitant microinjections of 5-HT_{1A}R and 5-HT_2R antagonists into the dorsal periaqueductal gray area, while impairment of escape is blocked by 5-HT_2R antagonist injection in this region [5]. In larval zebrafish, pharmacogenetic ablation of habenular neurons lead to freezing after fear conditioning, instead of the usual modification of freezing to flight [153] and impairs the acquisition of active avoidance [154]. Lesions of the fasciculus retroflexus, the efferent projection of the habenulae, are anxiogenic when made in neonate rats, but not adolescent animals; however, the effects of these lesions are diminished by social isolation before testing [155].

Although the habenula also presents extensive projections to the raphe nuclei [156], the role of these projections in the control of the serotonergic system is understudied. Stimulation of the LHb produces frequency-dependent effects; high frequency stimulation decreases firing of serotonergic neurons in the DRN, while low frequency stimulation produces the opposite effect [157–159]. Ferraro et al. [158] observed that low frequency LHb stimulation excites all raphe neurons studied, while high frequency LHb stimulation inhibited only serotonergic cells, an effect which is antagonized by bicuculline. Therefore, glutamatergic LHb projections synapse on at least two neuron classes in the raphe, namely serotonergic and putatively GABAergic cells. LHb also sends substance P-ergic afferents to the dorsal raphe nucleus [160], and local application of substance P in the DRN increases the frequency of spontaneous excitatory postsynaptic currents in serotonergic neurons [161] and increases serotonin efflux in the DRN and prefrontal cortex (but not ventral hippocampus) [162], an effect which is blocked by AMPA/kainate antagonists, suggesting that NK_1Rs are to be found in glutamatergic neurons. The efflux on the DRN and prefrontal cortex is dependent on $5\text{-}HT_{1A}$ receptors, as $5\text{-}HT_{1A}R$ knockout mice and animals treated with WAY 100635 do not present this effect [162].

3.6 Nodal Structures Regulating Panic: The Cerebral Aversive System

While the medial prefrontal cortex, the extended amygdala, the ventral hippocampus, and the lateral habenula all represent structures which are part of a "behavioral inhibition system" [9], a central amygdaloid-medial hypothalamic-mesopontine "cerebral aversive system" has been proposed [15, 35, 163–166]. This circuit involves the control of passive and active defensive behaviors associated with distal and proximal threats. Much more is known about organizational principles of this modular, lateralized, hierarchical system than about the behavioral inhibition system.

Projections from each of these structures are topographically organized [156, 167–170], constituting veritable "modules" controlling passive and active behavioral strategies [171–173]. Thus, caudodorsal medial prefrontal cortical cell groups which project preferentially to the dorsolateral periaqueductal gray area (dlPAG) also project selectively to the dorsomedial subdivision of the ventromedial hypothalamic nucleus (VMHdm) and to the anterior hypothalamic nucleus (AHA), and these regions show reciprocal connections with the dlPAG [164, 167, 169, 174, 175]. Similarly, orbitoinsular and rostroventral medial prefrontal cortical cell groups which project to the ventrolateral periaqueductal gray area (vlPAG) project selective to the lateral hypothalamus (LHA), which in its turn also shows reciprocal connections with the vlPAG [156, 166, 168–171, 173, 174, 176–184]. This dorsal versus ventral organization is observed, for example, in mice previously exposed to aversive stimuli, for example, a dorsal stream of c-Fos-like immunoreactivity is observed in the forebrain of animals which predominantly try to escape the stimulus, while a ventral stream is observed in animals which predominantly freeze [35]. The organization of defensive behavior in the cerebral aversive system is hierarchical: defensive behavior elicited by amygdalar or hypothalamic electrical stimulation can be prevented by periaqueductal gray lesions, while defensive responses elicited by PAG stimulation can still be elicited after lesions of the hypothalamus or amygdala [185, 186].

There is some evidence for functional hemispheric lateralization of the cerebral aversive system. Human carriers of the T polymorphism of the gene encoding tryptophan hydroxylase 2 show increased right amygdalar activation after processing of angry or fearful faces [187]. Kindling of the right CeA increases anxiety-like behavior in the elevated plus-maze one week after the last seizure [68]. In cats, a lateralized circuit comprising the projection from CeA to ventromedial hypothalamus is associated with defensive behavior, and FG-7142 induces long-term potentiation in the right CeA-VMH projection (but not in the left) and increases defensive behavior [188]. Rabbits exposed to predators show increased firing of right CeA neurons in relation to left CeA neurons [189]. Likewise, predator stress increases potentials evoked in the right (but not left) lateral periaqueductal gray area (lPAG) after central amygdala (CeA) stimulation [81, 190], and increases phosphorylation of cAMP response element binding protein (CREB) in the right lPAG [190]. Likewise, increasing CREB levels in the right (but not left) lPAG produces long lasting increases in anxiety-like behavior and CeA-lPAG transmission [191]. These effects are thought to be NMDAR-dependent, and associated with GABAergic tone on the right amygdalo-periaqueductal pathway [80, 81, 188, 192, 193]. Overall, these results suggest a right-lateralized circuit that mediates fear responses and residual (contextual) anxiety responses after predator stress; increased VH-BLA connectivity leads to disinhibition of the CeA efferent pathway to the lateral PAG and VMH, releasing defensive responses in these regions. These circuits are preferentially sensitized by anxiogenic treatments, such as administration of FG-7142, kindling or predator stress.

3.7 The Central Amygdala

The central nucleus of the amygdala (CeA) is a structure which has been implicated in fear responses [15]. This region receives important indirect projections from the lateral (LA) and basolateral (BLA) portions of the amygdala, forming pathways that are involved in fear conditioning [66]. These pathways are trisynaptic, involving a glutamatergic projection from the LA to a feed-forward inhibition microcircuit in the intercalated masses (ITC), which then disinhibits GABAergic projection neurons of the CeA [62, 194, 195]. ITC neurons show a high expression of OCT3, and thus extracellular monoamine concentrations in this region is expected to be higher after stressful events [196]. CeA-projecting neurons also receive massive projections from the infralimbic cortex [25–27], and IL stimulation reduces the excitability of brainstem-projecting CeA neurons [197].

The CeA can be roughly divided in three cell columns, a lateral capsular "shell", a lateral "core", and a medial "pallidal" region [200]. The pallidal CeA is activated during fear-potentiated startle, while both the pallidal CeA and the CeA shell are activated during fear conditioning [201]; strikingly, pallidal CeA activation during fear-potentiated startle is correlated with activation of neurons in the dorsal subregion of the dorsal raphe nucleus [201]. CeA columns can also be differentiated from the neurochemical point of view; for example, oxytocin receptors are located in neurons in the shell and core subregions, and the application of TGOT, an oxytocin agonist, depolarizes neurons in these columns [202]. Neurons which receive projection from these OXTR-expressing cells are inhibited by bath application of TGOT, while bath application of vasopressin excites these targets, which are located in the "pallidal" CeM [202]. Furthermore, optogenetic stimulation of CeM neurons or local inhibition of core and shell cells by muscimol induces unconditioned freezing [203], and optogenetic stimulation of BLA afferents in the core and/or shell cells (but not of the BLA per se) increases exploration of the open arms of an elevated plus-maze, while optogenetic inhibition of the same projections produce an opposite effect [16]. Microinjection of muscimol into the CeA core and shell before discriminative fear conditioning decreases freezing, 24 h later, to the conditioned stimulus that was associated with electrical shock; when muscimol is injected into the CeM between training and testing, conditioned freezing decreases [203].

Cells in the CeA core can be further subdivided in relation to their responses to a conditioned stimulus (CS^D) after discriminative fear conditioning [203]. About one-third of CeA core cells excited by the CS^D ("ON"-cells), are either CRFergic or dynorphinergic, and do not express the δ isoform of protein kinase C (PKCδ) [204]. One-fourth of these cells are inhibited by the CS^D ("OFF"-cells), express PKCδ and the oxytocin receptor, and are distinct from the CRFergic ON-cells [204]. The first cells are activated prior to the inhibition of OFF-cells, suggesting that the inhibitory projection from ON- to OFF-cells drives the inhibition of the latter by the CS^D [203]. This inhibition actually leads to the disinhibition of CeM output neurons, releasing freezing behavior that is tonically restrained by shell and core cells [203]. Consistently, pharmacogenetic inhibition of OFF-cells increases conditioned freezing to CS^D and post-CS freezing [204].

The CeA shows increased c-Fos-like immunoreactivity [205] and *egr*-1 mRNA expression [206] after contextual fear conditioning, and these effects are *increased* in a dose-dependent fashion after diazepam administration [206, 207]. The same is not observed after restraint stress, in which diazepam does not affect c-Fos-like immunoreactivity in the CeA, but dexamethasone decreases it [208]. In contrast, light-shock conditioning *decreases c-fos* mRNA expression in the CeA [209]. Fear conditioning leads to an increase in the firing rate of CeM neurons during CS presentation followed by a return to near-baseline levels 24 h later; after conditioning, CeA core cells respond at equal proportions to the CS with excitatory or inhibitory responses; 24 h later, the number of cells which are inhibited by the CS is tripled, with no changes in cells which are excited by the CS [210]. Thus, plasticity in CeA neurons is important in the acquisition and consolidation of conditioned fear responses, but these cells also respond to systemic, unconditioned stressful manipulations such as restraint stress [208], bacterial lipopolysaccharides [211], and acute treatment with serotonin reuptake blockers [212–214]. Consistent with this idea, Holahan and White [215] observed that rats exposed to fear conditioning show increased c-Fos-like immunoreactivity in the CeA whether or not they freeze, suggesting that amygdala neurons are activated by conditioned and unconditioned aversive events and not freezing. Interestingly, CeA CRFergic neurons are also activated after exposure of to an open field, an effect which is abolished in 5-HT_{2C}R knockouts [216]; however, the administration of mCPP into the CeA does not produce anxiogenic effects in the open field [96].

Chemical lesions of the CeA using ibotenic acid produce a variety of behavioral effects. Bilateral lesions increase the aversion toward quinine (a non-palatable drink) in rats, without effects on the acquisition of conditioned taste aversion [217]. Likewise, ibotenic acid lesions decrease postshock freezing and ultrasonic vocalizations during CS-US pairing in fear conditioning, but does not affect conditioned freezing 24 h after pairing [218]. However, electrolytic and ibotenic acid lesions of the CeA block fear-potentiated startle to visual or auditory CSs [219], an effect which is also observed with non-NMDAR antagonists [220]. It is possible that projection-specific impairments are involved in the discrepancy effects of CeA lesions on conditioned freezing and fear-potentiated startle; for example, projections from the rostral CeM arrive in the nucleus reticularis pontis caudalis [221], an important premotor structure in the startle pathway, while a different set of cells project to the periaqueductal gray or to the BNST [222], and yet another set projects to the dorsal vagal complex [223]. Nonetheless, inactivation of the CeA has also been demonstrated to impair unconditioned defensive responses. In rhesus monkeys, ibotenic acid lesions of the CeA impairs freezing in response to human invaders, and decreases the latency to retrieve palatable food items in the presence of snakes [224]. In rats, microinjections of muscimol into the CeA produce anxiolytic-like effects in the EPM [225]. Overall, it appears as if lesions of the CeA impair the ability to detect or discriminate aversive stimuli of different modalities and classes.

Serotonergic innervation is moderate in the shell and pallidal CeA, and scarce in the CeA core [226], and originates mainly in the dorsal (DRD) and caudal (DRC) subregions of the dorsal raphe nucleus [84, 85, 222, 227, 228]. Mice show a rostrocaudal gradient of overlap between serotonergic and CRFergic fibers in the CeA shell [226]. A class of GABAergic interneurons ("Type I" cells) in the CeA shell is characterized by a prominent depolarizing after-potential [131]. Activation of 5-HT_{1A} receptors in these cells decreases their spontaneous firing, releasing activity of GABAergic projection "Type II" neurons located in the CeA core [198](Figure X). These latter neurons are inhibited by CRF and excited by oxytocin [202, 226], being thus equivalent to the OFF-cells described above [203, 204]. Their activation inhibits vasopressin receptor expressing neurons in the pallidal CeA which project to the ventrolateral periaqueductal gray area, while at the same time disinhibiting cholinergic neurons in the nucleus basalis of Meynert and substantia innominata. The net result is a shift from freezing to risk assessment behavior and cortical arousal [198], placing the CeA core as the "switch" that is sensitive to serotonin (leading to risk assessment) or CRF (leading to freezing). A second, parallel set of neurons from the CeM does not respond to the application of oxytocin agonists on the CeA core, receiving direct projections from GABAergic neurons in the intercalated masses and projecting to the dorsal vagal complex [223]. Different from the CeM output neurons that control freezing, inhibition of these latter cells do not affect freezing, but instead decrease the bradycardia elicited by exposure to a context which has been associated with electrical shocks [223].

3.8 The Medial Hypothalamic Defense System

The hypothalamus has been recognized as a fundamental structure in the control of defensive behavior since at least the work of Bard [229]. Electrical or chemical stimulation work demonstrated that the medial zone of the hypothalamus integrates defensive responses [230–240]. Defensive responses elicited by stimulation of medial hypothalamic structures is different from that observed by stimulation of the dorsolateral PAG: while stimulation of the hypothalamus produces a directed escape response, PAG-evoked responses are characterized as sudden running/galloping bouts (akin to the "activity bout" observed in the first stages of fear conditioning) and aimless vertical jumps [237, 238]. Likewise, chemical lesions of medial hypothalamic structures impair the expression of defensive behavior in face of a predator [241], and this exposure upregulates c-Fos expression in the anterior hypothalamic nucleus (AHN), dorsomedial subregion of the ventromedial hypothalamus (VMHdm), dorsal premammillary nucleus (PMd), and paraventricular nucleus (PVN) [241, 242]. Nuclei from the medial hypothalamus are also activated after open field exposure [72], dlPAG stimulation at escape and freezing thresholds [243], and in mice exposed to ultrasonic stimulus in the home cage after sensitization by footshock (but not animals which have not been previously sensitized, and thus show more escape than freezing responses) [35]. c-Fos expression in the AHA is greater in high-anxiety (HAB) rats than in low-anxiety (LAB) rats exposed to an open field, the open arm of an elevated plus-maze, air-jet stress, and administration of FG-7142 [244]. Cytochrome oxidase activity in the ventromedial hypothalamus *decreases* after chronic social defeat stress in highly reactive rats, an effect which is also observed in the dlPAG [245]. The VMH is more activated in rats which have been exposed only once to cat odor, while the PMd is activated after one or two exposures or after exposure to a context which has been paired with a predator odor [246]. Microinjections of muscimol in the PMd interfere with unconditioned responses to predators and to predator-paired context, and dramatically decreasesc-Fos-like immunoreactivity in the rostromedial portions of the dlPAG in response to predatory context [247].

An important region in this regard is the dorsomedial hypothalamus (DMH) [248], which is activated by intraperitoneal injections of caffeine, yohimbine or FG-7142, but not the non-selective 5-HT_2 receptor agonist mCPP [249]. This region presents a high density of organic cation transporters from the OCT-3 type [196] and of plasma membrane monoamine transporters (PMAT) [250]. Restraint stress increases the content of serotonin, dopamine, and noradrenaline in the DMH of female rats [251], an effect which is potentiated by the uptake$_2$ blocker decynium-22 [252, 253] and is mimicked by intracerebroventricular corticosterone or CRF injections in roughskin newts [254] and Chinook salmons [255]; in this latter case, fluoxetine potentiates the effect of CRF, and the 5-HT$_{1A}$R antagonist NAN-190 blocks it. Decynium-22 injections in the DMH also increase spontaneous non-ambulatory activity in the home cage [252]. Chronic inhibition of GABA synthesis in the DMH leads to panic-like responses and cardiorespiratory arousal following infusions of hypertonic sodium lactate [256] or hypertonic sodium chloride but not iso-osmolar mannitol [257]. These responses are followed by decreased c-Fos immunoreactivity in serotonergic neurons located in the lateral wings of the dorsal raphe (lwDR) [258].

An important link between the dorsomedial hypothalamus and the hypothalamic nuclei which coordinate neuroendocrine responses to stress is observed by the fact that DMH injections of muscimol decrease c-Fos-like immunoreactivity in the parvocellular and magnocellular neurons of the paraventricular nucleus [259] and decreases changes in heart rate, blood pressure, and plasma ACTH secretion after air-jet stress [260, 261]. Magnocellular neurons of the PVN synthesize and secrete the nonapeptides vasopressin (AVP) and oxytocin (OXT), which are secreted into the hypothalamus after stressful situations [262–264] and which regulate plasma corticosteroid levels [265]. Likewise, parvocellular neurons from the PVN synthesize and release AVP and CRF into the long portal blood vessels, synergistically stimulating ACTH secretion from the corticotrope cells of the anterior pituitary [266]. Stress-induced activation of parvocellular PVN neurons is blocked by nitric oxide synthase inhibitors, demonstrating that this pathway is NO-dependent [267]; nonetheless, NOS inhibitors do not block the rapid non-genomic inhibition of glutamate release by corticosteroids, an effect which is mediated by endocannabinoids released from the postsynaptic parvocellular neuron into a presynaptic glutamatergic cell [268].

Serotonin receptors in the dorsomedial hypothalamus were also implicated in the regulation of neuroendocrine responses to stress [269]. Mice knocked out of the gene encoding SERT show increased sensitivity to stress [270, 271] and reduced density of 5-HT$_{1A}$Rs in the hypothalamus [272]. Restoring 5-HT$_{1A}$R levels in the medial hypothalamus of these animals eliminates the exaggerated plasmatic release of ACTH to saline injection, but did not change anxiety-like behavior in the elevated plus-maze [273]. Conversely, reducing the density of 5-HT$_{1A}$ receptors in the DMH, VMH, PVN, and AHN in SERT$^{+/+}$ mice does not alter behavior in the EPM [273]. Treatment with DOI, an agonist at 5-HT$_{2A}$ and 5-HT$_{2C}$ receptors, increases plasma levels of ACTH, corticosterone, oxytocin, prolactin, and renin, and activates CRFergic and oxytocinergic (but not vasopressinergic) neurons in the PVN [274]. It seems that medial hypothalamic 5-HT$_{1A}$Rs can regulate neuroendocrine responses to stress, but not anxiety-like behavior.

3.9 The Mesopontine Rostromedial Tegmental Nucleus

The rostromedial tegmental nucleus (RMTg) is a GABAergic structure which receives a major input from the lateral habenula and extended amygdala, and heavily projects to the all subregions of the dorsal raphe nucleus, the substantia nigra and ventral tegmental area, the ventrolateral PAG, the dorsal hypothalamus, and dorsal hippocampus [28, 275]. This region can be divided in a core and a shell subregion, with the first presenting the higher density of GABAergic cells [28]. VTA-projecting LHb afferent-recipient neurons in RMTg core are activated by footshocks in rats and inhibited after feeding in food-deprived animals. These neurons also respond at short latency to stimuli which are predictive of shock or sucrose delivery, with the majority of cells responding more to shock-predictive cues than to sucrose-predictive cues. A negative gradient of aversive preference in cell firing rates is observed moving from the core to the shell. Shock cue-preferring neurons are valence-consistent in that they also respond maximally to shock US, while sucrose cue-preferring neurons are valence-reverted in that they respond maximally to shock US or are inhibited by sucrose US. A small proportion of cells ($\sim$25 %) respond maximally when USs (either sucrose or shock) was omitted. Finally, RMTg lesions markedly reduce conditioned freezing to an auditory CS and unconditioned freezing to predator odor, while increasing burying/treading to odor and abolishing predator-odor-induced locomotor suppression and increasing the time spent in the open arms of an elevated plus-maze [174]. These effects are opposite to what is observed after habenula lesions in rats and zebrafish [4, 5, 153–155].

3.10 The Periaqueductal Gray Area

The periaqueductal gray area (PAG) is a "canonical" structure in the cerebral aversive system [170, 171]. This region presents a columnar organization, with inputs and outputs segregated in a dorsoventral and mediolateral pattern [171]. A dorsolateral (dlPAG) zone stains heavily for nitric oxide synthase [170], while the ventrolateral (vlPAG) zone presents intense staining for phenylethanolamine *N*-methyltransferase (the enzyme which converts norepinephrine into epinephrine) [276], connexins 30 and 43 [277], and calcitonin gene-related peptide [278]. Classically, the ventral portion is thought to mediate "passive" defense reactions (freezing/tonic immobility, opioid-mediated analgesia, and hypotension and bradycardia), while the dorsal portion mediates "active" defense reactions (escape, non-opioid-mediated analgesia, and hypertension and tachycardia)[170]. For example, "inescapable" noxious stimuli selectively activate the vlPAG, while "escapable" noxious stimuli activate the dlPAG [279]. Escape, but not inhibitory avoidance, performance in the elevated T-maze increases c-Fos-like immunoreactivity in the dorsolateral, but not ventrolateral, PAG [280], and the dlPAG, but not vlPAG, is activated after electrical stimulation of the inferior colliculus at escape (but not freezing) threshold [75]. Likewise, animals exposed to predators show dlPAG activation, while animals exposed to swim stress show vlPAG activation [281]. After repeated inescapable stress, ΔFosB is upregulated in the vlPAG, but not in the dlPAG [282]. These results point to an important role of the vlPAG in the response to inescapable stress, consistent with the idea that this column mediates "passive" coping. In guinea pigs, although both columns show increases in c-Fos-like immunoreactivity after induction of tonic immobility, the increase observed in the ventrolateral PAG is greater [283]; in rats, contextual fear conditioning increases Fos-like immunoreactivity more in the vlPAG than in the dlPAG [284], and anxiogenic drugs (yohimbine, mCPP, caffeine, and BOC-CCK_4) produce a similar trend [117]. The same pattern is observed in mice which were withdrawn from chronic morphine; in this case, about 10 % of activated cells were GABAergic [285]. Exposure to cat odor induces c-Fos-like immunoreactivity in both PAG columns that is not reduced by systemic midazolam [242, 286], but predator stress increases CREB phosphorylation in the right dlPAG [190] and *egr-1* mRNA expression in the right vlPAG [287]. Likewise, the ventrolateral PAG shows Fos-like immunoreactivity after 'trial 1' and 'trial 2' in the elevated plus-maze as well as after exposure to the context in which animals were exposed to cat odor, while the dorsolateral PAG is more activated by 'trial 1' than 'trial 2' and not activated by context at all [246]. About 24 h three exposures to cat odor, ΔFosB was upregulated in the ventrolateral, but not dorsolateral, column of the PAG [288]; it has been suggested that this "residual" activation is related to a role for vlPAG in stress recovery. Thus, acute, escapable stress induces rapid and transient c-Fos-like immunoreactivity in the dlPAG, and extension of this stress leads to the expression of more stable transcription factors (Erg-1, ΔFosB) in the vlPAG.

Dorsolateral PAG stimulation with stepwise increases in the electrical current elicits orienting responses, followed by freezing and then vigorous escape reactions [289–291]; ventral PAG stimulation immediately produces quiescence and tonic immobility [292–294]. Stimulation of ventral PAG leads to freezing that disappears as soon as the stimulation is turned off, while freezing elicited by dlPAG stimulation is sustained (post-stimulation freezing) [295]. Likewise, lesions of the vlPAG reduce conditioned freezing under distributed delayed shocks (which was argued to relate with post-encounter behavior), while lesions of the dlPAG enhance conditioned freezing under massed immediate shocks (which was argued to represent circa-strike behaviors) [296]. Moreover, contextual fear conditioning actually increases the current threshold to elicit escape behavior from dlPAG stimulation [297]. The capacity of vlPAG lesions to reduce conditioned freezing does not generalize to unconditioned freezing induced by dlPAG stimulation [298] suggesting different outputs for these two types of immobility.

The intrinsic circuitry of the dorsolateral and ventrolateral columns of the PAG has been proposed on the basis of pharmacological experiments (Fig. 3.3). In the dlPAG, GABAergic neurons receive information from the CeA, being inhibited by this projection; these "OFF-cells" are responsible for the suppression of aversive transmission, and are excited by activation of 5-HT_2 receptors. In this region, GABA is co-localized with 5-HT_{2A} receptors [299] and NADPH-dependent

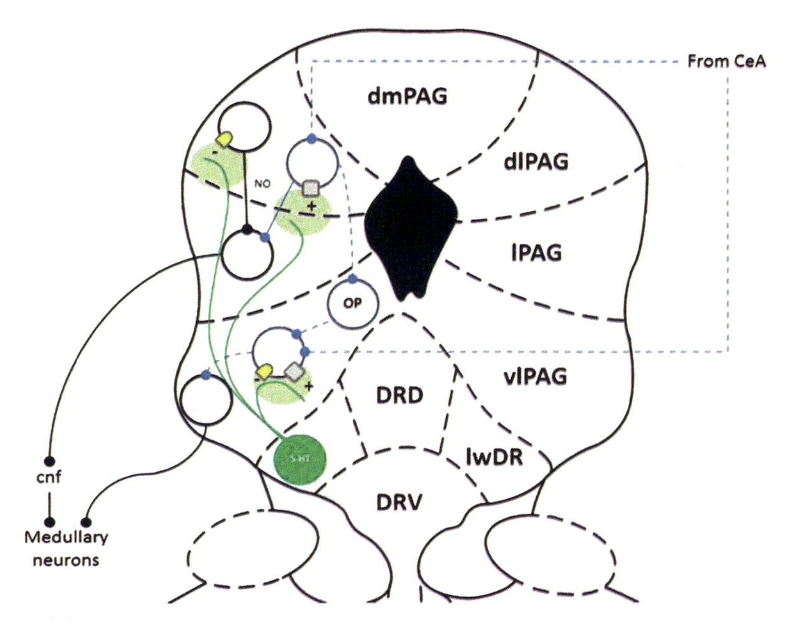

Fig. 3.3 Circuitry of the periaqueductal gray area (PAG) associated with defensive behavior. In the dorsolateral column (dlPAG), glutamatergic ON-cells are excited by aversive stimuli, while OFF-cells are inhibited by these stimuli. Both intrinsic neurons project to the output cells of the dlPAG. ON-cells are excited by 5-HT_{1A} agonists, and OFF-cells are inhibited by 5-HT_{2A} agonists. As a result, the activation of both receptor types by serotonin (released at varicosities from neurons projecting from the lateral wings of the dorsal raphe [lwDR]) contributes to inhibit the expression of fear. In the ventrolateral column (vlPAG), an opioidergic neuron inhibits the activity of a GABAergic interneuron which expresses both 5-HT_{1A}Rs (*yellow shapes*) and 5-HT_{2A}Rs (*gray rectangles*). Here, serotonin can lead to inhibition (via 5-HT_{1A}Rs) or excitation (via 5-HT_{2A}Rs). These GABAergic interneurons inhibit glutamatergic projection neurons from the vlPAG. Incoming information from the CeA inhibits vlPAG GABAergic interneurons and dlPAG OFF-cells, releasing inhibition of the opioidergic neuron. Blue dashed lines, inhibitory projections; black filled lines, excitatory projections. Abbreviations: *5-HT* 5-hydroxytryptamine, serotonin, *CeA* central amygdala, *cnf* cuneiform nucleus, *dlPAG* dorsolateral column of the periaqueductal gray area, *dmPAG* dorsomedial column of the periaqueductal gray area. *DRD*, dorsal subregion of the dorsal raphe nucleus, *DRV*, ventral subregion of the dorsal raphe nucleus, *lPAG* lateral column of the periaqueductal gray area, *lwDR* lateral wing of the dorsal raphe nucleus, *NO* nitric oxide, *OP* opioidergic neuron, *vlPAG* ventrolateral column of the periaqueductal gray area. Anatomic diagram from Paxinos and Watson [199]

diaphorase (NADPHd, a marker of nitric oxide synthase activity) [300]. The injection of bicuculline or FG-7142 into the dlPAG is panicogenic while injection of benzodiazepines is panicolytic [301]. Midazolam is anxiolytic and FG-7142 is anxiogenic in the EPM [302, 303], and blockade of glutamic acid decarboxylase (the rate-limiting enzyme in GABA synthesis) decreases freezing and escapes thresholds of electrical dlPAG stimulation [304]. Intra-dPAG microinjections of the 5-HT_2 antagonist ketanserin decreases the current threshold for freezing when animals are first exposed to fear conditioning and dPAG electrical stimulation is made on the same context (contextual freezing), but not in a different context or in animals which did not go through fear conditioning [305]. Moreover, ketanserin in the dPAG is not effective in the elevated T-maze [306, 307]. When animals which had their dPAG stimulated at escape thresholds are returned to the stimulation cage after 24 h, contextual freezing is not blocked by ketanserin, but contextual antinociception is [308]. In addition, intra-dPAG ketanserin injections increase the duration of tonic immobility in guinea pigs [309]. Consistent with these observations, the administration of DOI (a 5-HT$_{2A}$/5-HT$_{2C}$ agonist) into the dPAG produces an anxiolytic-like effect in the elevated plus-maze in maze-experienced rats ('trial 2'), while 5-HT$_{2C}$ agonists produces anxiolytic effects in both trial 1 and trial 2 [310, 311]. DOI and midazolam also block the aversive effects of intra-dlPAG administration of the nitric oxide donor SIN-1 [312].

The "switch" mechanism proposed for the CeA (Fig. 3.2) thus negatively regulates dlPAG OFF-cells. When the switch is "on" (i.e., when Type I neurons are not inhibited by activation of 5-HT$_{1A}$Rs), dlPAG OFF-cells will be inhibited by the GABAergic projection from the CeM. In this case, the activity of glutamatergic projection neurons from the dlPAG will be released, leading to fear (i.e., freezing/ escape reactions). When the switch is "off" (i.e., when Type I neurons are inhibited), dlPAG OFF-cells are disinhibited, leading to anxiety (i.e., risk assessment).

A second pool of neurons is composed of glutamatergic cells which are excited by aversive stimuli [313]. These "ON-cells" present 5-HT$_{1A}$ and P2X purinergic receptors [314], and intra-dlPAG administration of 8-OH-DPAT depresses neuronal firing [315] and attenuates escape behavior induced by the local injection of homocysteic acid [316] and electrical stimulation in this region [317], but not escape induced by the nitric oxide donor SIN-1 [312, 318]. Similarly, freezing to context and inhibitory avoidance acquisition are impaired after 8-OH-DPAT microinjections [307, 317], but not WAY 100635 [306, 307], on the dlPAG. Likewise, releasing inhibition on serotonergic neurons of the lateral wing of the DRN by microinjection of WAY 100635 or kainate blocks the escape reaction induced by homocysteic acid, but not SIN-1, in the dlPAG [319]. These results strongly suggest that NO in the dlPAG affects mainly the first pool of neurons, without affecting 5-HT$_{1A}$R-positive cells.

The circuit delineated in Fig. 3.3 for the dlPAG was proposed by analogy with a vlPAG-rostroventrolateral medulla (RVLM) projection composed on "ON" and "OFF"-cells [320]. RVLM OFF-cells show an abrupt pause in firing just before the occurrence of nocifensive reflexes, and are continuously active during systemic or local morphine administration into the RVLM, PAG, or amygdala [320–323]. ON-cells show a burst of activity just before a nocifensive reflex, and this activity is depressed during systemic or local morphine administration into the RVLM, PAG, or amygdala [321–325]. Neutral cells do not respond to nociceptive stimuli, and its activity is not directly related to nocifensive reflexes [320]. These observations highlight another important component of the defensive reaction that is gated by the PAG, viz, fear-induced analgesia [326]. Considering the properties of GABAergic neurons in the vlPAG and the inhibitory effect of opioids in this region, Behbehani [327] proposed a model in which PAG neurons that project to the RVLM are under tonic inhibition by GABAergic interneurons; opioid agonists inhibit these interneurons, releasing PAG projection cells that cause activation of OFF-cells in the RVLM.

These neurons are also under the influence of serotonin. Administration of the nonspecific 5-HT antagonist methysergide into the vlPAG inhibits the antinociceptive effect produced by electrical stimulation [328] or by microinjection of morphine in this structure [329, 330], as well as by central (but not medial) amygdala stimulation [331]. 5-HT1A receptors in the vlPAG have been implicated in the control of anxiety, but not fear, since microinjection of both 8-OH-DPAT and WAY 100635 are anxiolytic in the elevated T-maze, but do not impair escape responses [332]. The effect of WAY 100635 suggests a tonic effect of serotonin at 5-HT1A receptors that is not observed in the dorsolateral PAG. 5-HT_2 receptors are also under a tonic effect, since microinjection of ketanserin blocks antinociception elicited by electrical stimulation of this structure at a current which elicits freezing in rats [330, 333]. However, freezing elicited by vlPAG stimulation is not blocked by ketanserin [333], while this drug increases the duration of tonic immobility in guinea pigs [309]. DOI, on the contrary, impairs inhibitory avoidance, but not escape performance, in the ETM when applied in the vlPAG of rats [332], suggesting that 5-HT_2 receptors have a participation in anxiety-like behavior, but not panic responses, associated with the vlPAG.

3.11 Locus Coeruleus

Neurons in the locus coeruleus (LC) are the principal source of norepinephrine in the central nervous system, and the sole source of norepinephrine (NE) to the cortex and hippocampus [334–336]. This system has been implicated in arousal, contributing to the induction and maintenance of arousal and modulating collection and processing of behaviorally salient sensory information [337]. This region is rich in astrocytes, which are electrotonically coupled to LC neurons [338]. Changes in electrotonic coupling in the LC are associated with changes in attentiveness in target detection tasks [339].

Projections to the LC are topographically organized, with the nucleus of the solitary tract (NTS) and the vlPAG preferentially targeting the rostroventromedial subregion, the CRFergic neurons from the BNST, CeA core, and PVN projecting to the rostrodorsolateral subregion, and glutamatergic and enkephalinergic neurons from the gigantocellular reticular nucleus projecting to the LC "core" [340–344]. Inhibition of LC neurons in the rostral poles is associated with reduced arousal and responsiveness [345]. Coerulear projections also show subtle topographically distinguishable gradients; hippocampal projections come from fusiform cells in the dorsal portions of the LC, while hypothalamic projections come from small round and multipolar cells on the rostral aspects [346–348]. Different LC pathways project to the dorsal and ventral portions of the dentate gyrus [349].

These target regions also respond differently to norepinephrine. Acute immobilization stress induces NE release in the central amygdala [350] and an α_1-adrenoceptor-dependent increase in anxiety-like behavior [351]. In CeA, NE levels after 30 min of training are highly correlated with retention of inhibitory avoidance [352]. In rats selected for superior versus inferior acquisition of two-way avoidance (Roman lines), these effects are divergent; the infusion of norepinephrine in the CeA of high-avoidant rats during retention leads to a shift toward active behavior (burying, sniffing) in recall, while the same manipulation leads to passive (immobility) behavior during recall in low-avoidant rats [353]. In the basolateral amygdala, microinjection of norepinephrine inhibits spontaneous and evoked responses, an effect which is potentiated by systemic administration of the β-adrenoceptor antagonist propranolol and diminished by the α_2-adrenoceptor antagonist yohimbine [354]; systemic administration of yohimbine alone doubles extracellular NE levels in the BLA (through blockade of autoreceptors in the LC) and increases spontaneous activity in the lateral and basomedial regions of the BLA [355]. Therefore, in the BLA α_2-adrenoceptors mediate the inhibitory action of norepinephrine.

Microinjections of norepinephrine in the vlPAG increase arterial pressure and decrease heart rate in unanesthetized rats, an effect which is blocked by AVP_1R antagonists and is accompanied by increases in circulating vasopressin levels [356]. Likewise, chronic, peripheral administration of maprotiline, a selective norepinephrine reuptake inhibitor, or clomipramine, a non-selective serotonin and norepinephrine reuptake inhibitor, increases the amount of current necessary to elicit immobility by dlPAG stimulation [291]. Since midazolam only attenuates freezing at sedative doses, Schenberg and colleagues proposed that immobility in this model is representative of an NE-mediated attentional response [291], in spite of the suggestions of a role of ascending NErgic projections in anxiety disorders [357, 358]. In humans, electrical stimulation of the locus coeruleus increases plasma concentration of norepinephrine metabolites and produces alertness, without accompanying reports of anxiety [359], supporting Schenberg's hypothesis.

Other evidence for a role of NE in arousal instead of anxiety comes from activation studies in different stressor contexts. Rats selected for high anxiety-like behavior in the elevated plus-maze show increased Fos-like immunoreactivity in the LC after air-jet stress and administration of the benzodiazepine receptor inverse agonist FG-7142 [32]; however, no difference in Fos expression was observed between high- and low-anxiety rats exposed to predators or to swim stress [281], and previous footshock (which decreases flight responses to ultrasonic stimulation and increases freezing) does not alter Fos expression in the LC in mice [35]. Overall, these results suggest that LC neurons are activated by stressors regardless of quality or passive-versus-active behavioral output.

Activation studies were also used to probe the role of the LC in response to anxiogenic drugs. The α_2-adrenoceptor antagonist yohimbine, a drug which potentiates flight responses to predators [360] and increases the firing rate of LC neurons [361], increases c-Fos-like immunoreactivity in the locus coeruleus of rats

[117]. Transgenic mice overexpressing the neurotrophin-3 (TrkC) receptor show increases in the number of noradrenergic neurons in the LC, increased expression of α_{2A}- and α_{2B}-adrenoceptors in the brainstem, and increased anxiety-like behavior and responsiveness to lactate [362, 363]; these animals do not differ from control littermates in c-Fos-like immunoreactivity after yohimbine administration in all brain regions except the hypothalamus, where immunoreactivity is lower in transgenic mice [363].

In addition to yohimbine, other anxiogenic drugs (e.g., FG-7142, mCPP, caffeine, BOC-CCK_4, acute fluoxetine) also produce a marked increase in Fos expression in the LC [117, 364, 365]. In the case of fluoxetine, acute administration at doses which produce an anxiogenic-like effect in the social interaction and air-jet escape tests selectively increases Fos-like immunoreactivity in the LC after stress, also doubling norepinephrine efflux in the medial prefrontal cortex after air-jet stress [364]. These responses are blocked by 5-HT_{2C}R antagonists (but not 5-HT_{1A}R, 5-HT_{1B}R, or 5-HT_3R antagonists) and mimicked by 5-HT_{2C}R agonists, suggesting that acute fluoxetine exerts its anxiogenic effects through activation of 5-HT_{2C} receptors in the LC [365].

The LC presents an important projection to raphe neurons, with the caudal portion of the LC core projecting to caudal and ventral DRN, and the midcaudal portion projecting to the lateral wings of the DRN [366]. In vitro, the release of serotonin in DRN slices is strongly inhibited by norepinephrine or α_2-adrenoceptor agonists, an effect which is inhibited by α_2-adrenoceptor antagonists [367]. In vivo, the application of α_2-adrenoceptor agonists decreases, while application of α_2-adrenoceptor antagonists enhances, 5-HT release in the DRN [368, 369]. Thus, norepinephrine exerts a tonic inhibition of serotonergic transmission via α_{2A}-adrenoceptors. The mechanism for this inhibition is not entirely clear, as α_{2A}-adrenoceptors are important autoreceptors which limit the release of norepinephrine. Nonetheless, a postsynaptic effect has been described in the caudal subregion of the DRN, where norepinephrine inhibits calcium currents (but not inwardly rectifying potassium currents) in serotonergic neurons through an α_{2A}-adrenoceptor-dependent mechanism [370].

The origin of the serotonergic innervation of the LC is still disputed [371]. Immunocytochemical, autoradiographic, and release experiments all have shown that the LC receives a dense serotonergic innervation [372–374]. Injection of retrograde tracers into the LC revealed that the dorsal subregion of the DRN sends collaterals to this region [85]; injection of anterograde tracers in the DRN reveals that these projections synapse onto dendrites in the pericoerulear region, an area of neuropil surrounding the LC core [341]. Serotonin release in the locus coeruleus is mediated by nitric oxide, as superfusion with NO donors and L-arginine increases 5-HT release in this region, and NOS inhibitors decrease excitatory amino acid- or stress-evoked increases in 5-HT release [375]. Serotonin in the LC is predominantly inhibitory [376], also being able to attenuate glutamate-evoked, but not acetylcholine-evoked, excitatory responses [345, 377, 378].

References

1. Davis M, Rainnie D, Cassell M (1994) Neurotransmission in the rat amygdala related to fear and anxiety. Trends Neurosci 17:208–214
2. Bannerman DM, Rawlins JNP, McHugh SB, Deacon RMJ, Yee BK, Bast T, Zhang W-N, Pothuizen HHJ, Feldon J (2004) Regional dissociations within the hippocampus—Memory and anxiety. Neurosci Biobehav Rev 28:273–283
3. Amat J, Sparks PD, Matus-Amat P, Griggs J, Watkins LR, Maier SF (2001) The role of the habenular complex in the elevation of dorsal raphe nucleus serotonin and the changes in the behavioral responses produced by uncontrollable stress 97:118–126
4. Pobbe RLH, Zangrossi H Jr (2008) Involvement of the lateral habenula in the regulation of generalized anxiety- and panic-related defensive responses in rats. Life Sci 82:1256–1261
5. Pobbe RLH, Zangrossi H Jr (2010) The lateral habenula regulates defensive behaviors through changes in 5-HT-mediated neurotransmission in the dorsal periaqueductal gray matter. Neurosci Lett 479:87–91
6. Radley JJ, Arias CM, Sawchenko PE (2006) Regional differentiation of the medial prefrontal cortex in regulating adaptive responses to acute emotional stress. J Neurosci 26:12967–12976
7. Sullivan RM, Gratton A (2002) Behavioral effects of excitotoxic lesions of the ventral medial prefrontal cortex in the rats are hemisphere-dependent. Brain Res 927:69–79
8. Adhikari A, Topiwala MA, Gordon JA (2010) Synchronized activity between the ventral hippocampus and the medial prefrontal cortex during anxiety. Neuron 65:257–269
9. Gray JA, McNaughton N (2000) Neuropsychology of anxiety: An enquiry into the functions of the septo-hippocampal system, 2nd edn. Oxford University Press, Oxford
10. McNaughton N, Corr PJ (2004) A two-dimensional neuropsychology of defense: Fear/anxiety and defensive distance. Neurosci Biobehav Rev 28:285–305
11. Hikosaka O (2010) The habenula: From stress evasion to value-based decision-making. Nat Rev Neurosci 11:503–513
12. Zvolensky MJ, Lejuez CW, Eifert GH (2000) Prediction and control: operational definitions for the experimental analysis of anxiety. Behav Res Ther 38:653–663
13. Mendl M, Burman OHP, Parker RMA, Paul ES (2009) Cognitive bias as an indicator of animal emotion and welfare: emerging evidence and underlying mechanisms. Appl Anim Behav Sci 118:161–181
14. Davidson TL, Jarrard LE (2004) The hippocampus and inhibitory learning: a 'Gray' area? Neurosci Biobehav Rev 28:261–271
15. Davis M, Shi C (1999) The extended amygdala: are the central nucleus of the amygdala and the bed nucleus of the stria terminalis differentially involved in fear versus anxiety? Ann N Y Acad Sci 877:281–291
16. Tye KM, Prakash R, Kim S-Y, Fenno LE, Grosenick L, Zarabi H, Thompson KR, Gradinaru V, Ramakrishnan C, Deisseroth K (2011) Amygdala circuitry mediating reversible and bidirectional control of anxiety. Nature 471:358–362
17. Vogt BA, Vogt L, Farber NB (2004) Cingulate cortex and disease models. In: Paxinos G (ed) The rat nervous system, 3rd edn. Elsevier, New York, pp 705–727
18. van Groen T, Kadish I, Wyss JM (1999) Efferent connections of the anteromedial nucleus of the thalamus of the rat. Brain Res Rev 30:1–26
19. Aggleton JP, Brown MW (1999) Episodic memory, amnesia, and the hippocampal-anterior thalamic axis. Behav Brain Sci 22:425–444
20. Sripanidulchai K, Sripanidulchai B, Wyss JM (1984) The cortical projection of the basolateral amygdaloid nucleus in the rat: a retrograde fluorescent dye study. J Comp Neurol 229:419–431
21. Jay TM, Glowinski J, Thierry AM (1989) Selectivity of the hippocampal projection to the prelimbic area of the prefrontal cortex in the rat. Brain Res 505:337–340

22. Vertes RP (2004) Differential projections of the infralimbic and prelimbic cortex in the rat. Synapse 51:32–58
23. Chiba T, Kayahara T, Nakano K (2001) Efferent projections of infralimbic and prelimbic areas of the medial prefrontal cortex in the Japanese monkey, Macaca fuscata. Brain Res 888:83–101
24. Takgishi M, Chiba T (1991) Efferent projections of the infralimbic (area 25) region of the medial prefrontal cortex in the rat: an anterograde tracer PHA-L study. Brain Res 566:26–39
25. Sesack SR, Deutch AY, Roth RH, Bunney BS (1989) Topographical organization of the efferent projections of the medial prefrontal cortex in the rat: An anterograde tract-tracing study with Phaseolus vulgaris leucoagglutinin. J Comp Neurol 290:213–242
26. Freedman LJ, Insel TR, Smith Y (2000) Subcortical projections of area 25 (subgenual cortex) of the macaque monkey. J Comp Neurol 421:172–188
27. McDonald AJ, Mascagni F, Guo L (1996) Projections of the medial and lateral prefrontal cortices to the amygdala: a Phaseolus vulgaris leucoagglutinin study in the rat. Neuroscience 71:55–75
28. Jhou TC, Geisler S, Marinelli M, DeGarmo BA, Zahm DS (2009) The mesopontine rostromedial tegmental nucleus: a structure targeted by the lateral habenula that projects to the ventral tegmental area of Tsai and substantia nigra compacta. J Comp Neurol 513:566–596
29. Nitschke JB, Sarinopoulos I, Oathes DJ, Johnstone T, Whalen PJ, Davidson RJ, Kalin NH (2009) Anticipatory activation in the amygdala and anterior cingulate in generalized anxiety disorder and prediction of treatment response. Am J Psychiatry 166:302–310
30. Straube T, Schmidt S, Weiss T, Mentzel H-J, Miltner WHR (2009) Dynamic activation of the anterior cingulate cortex during anticipatory anxiety. NeuroImage 44:975–981
31. Simpson JR, Jr, Drevets WC, Snyder AZ, Gusnard DA, Raichle ME (2000) Emotion-induced changes in human medial prefrontal cortex: II. During anticipatory anxiety. Proceedings of the National Academy of Sciences USA, vol 98, pp 688–693
32. Salchner P, Sartori SB, Sinner C, Wigger A, Frank E, Landgraf R, Singewald GM (2006) Airjet and FG-7142-induced Fos expression in rats selectively bred for high and low anxiety-related behavior. Neuropharmacology 50:1048–1058
33. Salomé N, Salchner P, Viltart O, Sequeira H, Wigger A, Landgraf R, Singewald N (2004) Neurobiological correlates of high (HAB) versus low anxiety-related behavior (LAB): differential Fos expression in HAB and LAB rats. Biol Psychiatry 55:715–723
34. Frank E, Salchner P, Aldag JM, Salomé N, Singewald N, Landgraf R, Wigger A (2006) Genetic predisposition to anxiety-related behavior determines coping style, neuroendocrine responses, and neuronal activation during social defeat. Behav Neurosci 120:60–71
35. Mongeau R, Miller GA, Chiang E, Anderson DJ (2003) Neural correlates of competing fear behaviors evoked by an innately aversive stimulus. J Neurosci 23:3855–3868
36. Stalnaker TA, España RA, Berridge CW (2009) Coping behavior causes asymmetric changes in neuronal activation in the prefrontal cortex and amygdala. Synapse 63:82–85
37. Berridge CW, Mitton E, Clark W, Roth RH (1999) Engagement in a non-escape (displacement) behavior elicits a selective and lateralized suppression of frontal cortical dopaminergic utilization in stress. Synapse 32:187–197
38. Lacroix L, Spinelli S, Heibreder CA, Feldon J (2000) Differential role of the medial and lateral prefrontal cortices in fear and anxiety. Behav Neurosci 114:1119–1130
39. Blanco E, Castilla-Ortega E, Miranda R, Begega A, Aguirre JA, Arias JL, Santín LJ (2009) Effects of medial prefrontal cortex lesions on anxiety-like behaviour in restrained and non-restrained rats. Behav Brain Res 201:338–342
40. Hajós M, Richards CD, Szekely AD, Sharp T (1998) An electrophysiological and neuroanatomical study of the medial prefrontal cortical projection to the midbrain raphe nuclei in the rat. Neuroscience 87:95–108
41. Amat J, Baratta MV, Paul E, Bland ST, Watkins LR, Maier SF (2005) Medial prefrontal cortex determines how stressor controllability affects behavior and dorsal raphe nucleus. Nat Neurosci 8:365–371

42. Amat J, Paul E, Zarza C, Watkins LR, Maier SF (2006) Previous experience with behavioral control over stress blocks the behavioral and dorsal raphe nucleus activating effects of later uncontrollable stress: role of the ventral medial prefrontal cortex. J Neurosci 20:13264–13272
43. Amat J, Paul E, Watkins LR, Maier SF (2008) Activation of the ventral medial prefrontal cortex during an uncontrollable stressor reproduces both the immediate and long-term protective effects of behavioral control. Neuroscience 154:1178–1186
44. Baratta MV, Zarza CM, Gomez DM, Campeau S, Watkins LR, Maier SF (2009) Selective activation of dorsal raphe nucleus-projecting neurons in the ventral medial prefrontal cortex by controllable stress. Eur J Neurosci 30:1111–1116
45. Lowry CA, Hollis JH, de Vries A, Pan B, Brunet LR, Hunt JRF, Paton JFR, van Kampen E, Knight DM, Evans AK, Rook GAW, Lightman SL (2007) Identification of an immune-responsive mesolimbocortical serotonergic system: potential role in regulation of emotional behavior. Neuroscience 146:756–772
46. Hollis JH, Evans AK, Bruce KPE, Lightman SL, Lowry CA (2006) Lipopolysaccharide has indomethacin-sensitive actions on Fos expression in topographically organized subpopulations of serotonergic neurons. Brain Behav Immun 20:569–577
47. Beekman M, Flachskamm C, Linthorst ACE (2005) Effects of exposure to a predator on behaviour and serotonergic neurotransmission in different brain regions of C57bl/6N mice. Eur J Neurosci 21:2825–2836
48. Lehner M, Taracha E, Skórzewska A, Maciejak P, Wisłowska-Stanek A, Zienowicz M, Szyndler J, Bidziński A, Płaźnik A (2006) Behavioral, immunocytochemical and biochemical studies in rats differing in their sensitivity to pain. Behav Brain Res 171:189–198
49. Rex A, Fink H (1998) Effects of cholecystokinin-receptor agonists on cortical 5-HT release in guinea pigs on the X-maze. Peptides 19:519–526
50. Rex A, Marsden CA, Fink H (1997) Cortical 5-HT-CCK interactions and anxiety-related behaviour of guinea-pigs: a microdialysis study. Neurosci Lett 228:79–82
51. Rex A, Fink H, Marsden CA (1994) Effects of BOC-CCK-4 and L 365.260 on cortical 5-HT release in guinea-pigs on exposure to the elevated plus maze. Neuropharmacology 33:559–565
52. Rex A, Marsden CA, Fink H (1993) Effect of diazepam on cortical 5-HT release and behaviour in the guinea-pig on exposure to the elevated plus-maze. Psychopharmacology 110:490–496
53. Puig MV, Watakabe A, Ushimaru M, Yamamori T, Kawaguchi Y (2010) Serotonin modulates fast-spiking interneuron and synchronous activity in the rat prefrontal cortex through 5-HT1A and 5-HT2A receptors. J Neurosci 30:2211–2222
54. Pum ME, Huston JP, Müller CP (2009) The role of cortical serotonin in anxiety and locomotor activity in Wistar rats. Behav Neurosci 123:449–454
55. Dracheva S, Lyddon R, Barley K, Marcus SM, Hurd YL, Byne WM (2009) Editing of serotonin 2C receptor mRNA in the prefrontal cortex characterizes high-novelty locomotor response behavioral trait. Neuropsychopharmacology 34:2237–2251
56. Zhang X, Beaulieu J-M, Sotnikova TD, Gainetdinov RR, Caron MG (2004) Tryptophan hydroxylase-2 controls brain serotonin synthesis. Science 305:217
57. Stiedl O, Radulovic J, Lohmann R, Birkenfeld K, Palve M, Kammermeir J, Sananbenesi F, Spiess J (1999) Strain and substrain differences in context- and tone-dependent fear conditioning of inbred mice. Behav Brain Res 104:1–12
58. Yilmazer-Hanke DM, Roskoden T, Zilles K, Schwegler H (2003) Anxiety-related behavior and densities of glutamate, GABAA, acetylcholine and serotonin receptors in the amygdala of seven inbred mouse strains. Behav Brain Res 145:145–159
59. Englander MT, Dulawa SC, Bhansali P, Schmauss C (2005) How stress and fluoxetine modulate serotonin 2C receptor pre-mRNA editing. J Neurosci 25:648–651
60. Aggleton JP (2000) The amygdala: a functional analysis. Oxford University Press, New York
61. Rauch SK, Shin LM, Wright CI (2003) Neuroimaging studies of amygdala function in anxiety disorders. Ann N Y Acad Sci 985:389–410
62. Paré D, Quirk GJ, LeDoux JE (2004) New vistas on amygdala networks in conditioned fear. Am J Physiol 92:1–9

63. LeDoux J (2007) The amygdala. Curr Biol 17:R868–R874
64. Adolphs R, Tranel D, Damasio H, Damasio AR (1995) Fear and the human amygdala. J Neurosci 15:5879–5891
65. de Olmos JS, Beltramino CA, Alheid G (2004) Amygdala and extended amygdala of the rat: a cytoarchitectonical, fibroarchitectonical, and chemoarchitectonical survey. In: Paxinos G (ed) The rat nervous system, 3rd edn. Elsevier, New York, pp 509–603
66. Cain CK, LeDoux JE (2008) Brain mechanisms of Pavlovian and instrumental aversive conditioning. In: Blanchard DC, Blanchard RJ, Griebel G, Nutt D (eds) Handbook of anxiety and fear. Elsevier B. V, Amsterdam, pp 103–124
67. Fuller TA, Russchen FT, Price JL (1987) Sources of presumptive glutamatergic/aspartergic afferents to the rat ventral striatopallidal region. J Comp Neurol 258:317–338
68. Adamec R, Shallow T (2000) Rodent anxiety and kindling of the central amygdala and nucleus basalis. Physiol Behav 70:177–187
69. Miller HP, Levey AI, Rothstein JD, Tzingounis A, Conn PJ (1997) Alterations in glutamate transporter protein levels in kindling-induced epilepsy. J Neurochem 68:1564–1570 V
70. Lee Y, Gaskins D, Anand A, Shekhar A (2007) Glial mechanisms in mood regulation: a novel model of mood disorders. Psychopharmacology 191:55–65
71. Hale MW, Bouwknecht JA, Spiga F, Shekhar A, Lowry CA (2006) Exposure to high- and low-light conditions in an open-field test of anxiety increases c-Fos expression in specific subdivisions of the rat basolateral amygdaloid complex. Brain Res Bull 71:174–182
72. Hale MW, Hay-Schmidt A, Mikkelsen JD, Poulsen B, Shekhar A, Lowry CA (2008) Exposure to an open-field arena increases c-Fos expression in a distributed anxiety-related system projecting to the basolateral amygdaloid complex. Neuroscience 155:659–672
73. Hale MW, Hay-Schmidt A, Mikkelsen J, Poulsen B, Bouwknecht JA, Evans AK, Stamper CE, Shekhar A, Lowry CA (2008) Exposure to an open-field arena increases c-Fos expression in a subpopulation of neurons in the dorsal raphe nucleus, including neurons projecting to the basolateral amygdaloid complex. Neuroscience 157:733–748
74. Hale MW, Johnson PL, Westerman AM, Abrams JK, Shekhar A, Lowry CA (2010) Multiple anxiogenic drugs recruit a parvalbumin-containing subpopulation of GABAergic interneurons in the basolateral amygdala. Prog Neuropsychopharmacol Biol Psychiatry 34:1285–1293
75. Lamprea MR, Cardenas FP, Vianna DM, Castilho VM, Cruz-Morales SE, Brandão ML (2002) The distribution of Fos immunoreactivity in rat brain following freezing and escape responses elicited by electrical stimulation of the inferior colliculus. Brain Res 950:186–194
76. Orman R, Stewart M (2007) Hemispheric differences in protein kinase C III in the rat amygdala: Baseline asymmetry and lateralized changes associated with cue and context in a classical fear conditioning paradigm. Neuroscience 144:797–807
77. Andersen S, Teicher M (1999) Serotonin laterality in amygdala predicts performance in the elevated plus maze in rats. NeuroReport 10:3497–3500
78. Coleman-Mesches K, McGaugh JL (1995) Differential effects of pretraining inactivation of the right or left amygdala onr etention of inhibitory avoidance training. Behav Neurosci 109:642–647
79. Coleman-Mesches K, McGaugh JL (1995) Muscimol injected into the right of left amygdaloid complex differentially affects retention performance following aversively motivated training. Brain Res 676:183–188
80. Adamec RE, Burton P, Shallow T, Budgell J (1999) Unilateral block of NMDA receptors in the amygdala prevents predator stress-induced lasting increases in anxiety-like behavior and unconditioned startle—Effective hemisphere depends on the behavior. Physiol Behav 65:739–751
81. Adamec R, Blundell J, Burton P (2005) Role of NMDA receptors in the lateralized potentiation of amygdala afferent and efferent neural transmission produced by predator stress. Physiol Behav 86:75–91

82. Adamec R, Shallow T, Burton P (2005) Anxiolytic and anxiogenic effects of kindling—Role of baseline anxiety and anatomical location of the kindling electrode in response to kindling of the right and left basolateral amygdala. Behav Brain Res 159:73–88
83. Kalynchuk LE, Pinel JPJ, Meaney MJ (2006) Serotonin receptor binding and mRNA expression in the hippocampus of fearful amygdala-kindled rats. Neurosci Lett 396:38–43
84. Abrams JK, Johnson PL, Hollis JH, Lowry CA (2004) Anatomic and functional topography of the dorsal raphe nucleus. Ann N Y Acad Sci 1018:46–57
85. Imai H, Steindler D, Kitai ST (1986) The organization of divergent axonal projections from the midbrain raphe nuclei in the rat. J Comp Neurol 243:363–380
86. Amat J, Matus-Amat P, Watkins LR, Maier SF (1998) Escapable and inescapable stress differentially alter extracellular levels of 5-HT in the basolateral amygdala of the rat. Brain Res 812:113–120
87. Amat J, Tamblyn JP, Paul ED, Bland ST, Amat P, Foster AC, Watkins LR, Maier SF (2004) Microinjection of urocortin 2 into the dorsal raphe nucleus activates serotonergic neurons and increases extracellular serotonin in the basolateral amygdala. Neuroscience 129:509–519
88. Rainnie DG (1999) Serotonergic modulation of neurotransmission in the rat basolateral amygdala. J Neurophysiol 82:69–85
89. Koyama S, Kubo C, Rhee JS, Akaike N (1999) Presynaptic serotonergic inhibition of GABAergic synaptic transmission in mechanically dissociated rat basolateral amygdala neurons. J Physiol 518:525–538
90. Jiang X, Xing G, Yang C, Verma A, Zhang L, Li H (2009) Stress impairs 5-HT2A receptor-mediated serotonergic facilitation of GABA release in juvenile rat basolateral amygdala. Neuropsychopharmacology 34:410–423
91. Stutzmann GE, McEwen BS, LeDoux JE (1998) Serotonin modulation of sensory inputs to the lateral amygdala: dependency on corticosterone. J Neurosci 18:9529–9538
92. Cheng L–L, Wang S-J, Gean P-W (1998) Serotonin depresses excitatory synaptic transmission and depolarization-evoked Ca2+ influx in rat basolateral amygdala via 5-HT1A receptors. Eur J Neurosci 10:2163–2172
93. Zangrossi H Jr, Viana MB, Graeff FG (1999) Anxiolytic effect of intra-amygdala injection of midazolam and 8-hydroxy-2-(di-n-propylamino)tetralin in the elevated T-maze. Eur J Pharmacol 369:267–270
94. Guimarães FS, Zangrossi H Jr, Del Ben CM, Graeff FG (2010) Serotonin in panic and anxiety disorders. In: Müller C, Jacobs B (eds) Handbook of behavioral neurobiology of serotonin. Elsevier B. V, Amsterdam, pp 667–685
95. Sommer W, Möller C, Wiklund L, Thorsell A, Rimondini R, Nissbrandt H, Heilig M (2001) Local 5,7-dihydroxytryptamine lesions of rat amygdala: Release of punished drinking, unaffected plus-maze behavior and ethanol consumption. Neuropsychopharmacology 24:430–440
96. Campbell BM, Merchant KM (2003) Serotonin 2C receptors within the basolateral amygdala induce acute fear-like responses in an open-field environment. Brain Res 993:1–9
97. Cruz APDM, Pinheiro G, Alves SH, Ferreira G, Mendes M, Faria L, Macedo CE, Landeira-Fernandez J (2005) Behavioral effects of systemically administered MK-212 are prevented by ritanserin microinfusion into the basolateral amygdala of rats exposed to the elevated plus-maze. Psychopharmacology 182:345–354
98. Overstreet DH, Knapp DJ, Angel RA, Navarro M, Breese GR (2006) Reduction in repeated ethanol-withdrawal-induced anxiety-like behavior by site-selective injections of 5-HT1A and 5-HT2C ligands. Psychopharmacology 187:1–12
99. Christianson JP, Ragole T, Amat J, Greenwood BN, Strong PV, Paul ED, Fleshner M, Watkins LR, Maier SF (2010) 5-hydroxytryptamine 2C receptors in the basolateral amygdala are involved in the expression of anxiety after uncontrollable traumatic stress. Biol Psychiatry 67:339–345
100. Hackler EA, Airey DC, Shannon CC, Sodhi MS, Sanders-Bush E (2006) 5-HT2C receptor RNA editing in the amygdala of C57BL/6J, DBA/2J, and BALB/cJ mice. Neurosci Res 55:96–104

101. Thompson CL, Pathak SD, Jeromin A, Ng LL, MacPherson CR, Mortrud MT, Cusick A, Riley ZL, Sunkin SM, Bernard A, Puchalski RB, Gage FH, Jones AR, Bajic VB, Hawrylycz MJ, Lein ES (2008) Genomic anatomy of the hippocampus. Neuron 60:1010–1021
102. Leonardo ED, Richardson-Jones JW, Sibille E, Kottman A, Hen R (2006) Molecular heterogeneity along the dorsal-ventral axis of the murine hippocampal CA1 field: a microarray analysis of gene expression. Neuroscience 137:177–186
103. Fanselow MS, Dong H-W (2010) Are the dorsal and ventral hippocampus functionally distinct structures? Neuron 65:7–19
104. Maggio N, Segal M (2007) Striking variations in corticosteroid modulation of long-term potentiation along the septotemporal axis of the hippocampus. J Neurosci 27:5757–5765
105. Maggio N, Segal M (2009) Differential modulation of long-term depression by acute stress in the rat dorsal and ventral hippocampus. J Neurosci 29:8633–8638
106. Royer S, Sirota A, Patel J, Buzsáki G (2010) Distinct representations and theta dynamics in dorsal and ventral hippocampus. J Neurosci 30:1777–1787
107. Steullet P, Cabungcal J-H, Kulak A, Kraftsik R, Chen Y, Dalton TP, Cuenod M, Do KQ (2010) Redox dysregulation affects the ventral but not dorsal hippocampus: impairment of parvalbumin neurons, gamma oscillations, and related behaviors. J Neurosci 30:2547–2558
108. Bertoglio LJ, Joca SRL, Guimarães FS (2006) Further evidence that anxiety and memory are regionally dissociated within the hippocampus. Behav Brain Res 175:183–188
109. Pentkowski NS, Blanchard DC, Lever C, Litvin Y, Blanchard RJ (2006) Effects of lesions to the dorsal and ventral hippocampus on defensive behaviors in rats. Eur J Neurosci 23: 2185–2196
110. Richmond MA, Yee BK, Pouzet B, Veenman L, Rawlins JNP, Feldon J, Bannerman DM (1999) Dissociating context and space within the hippocampus: effects of complete, dorsal and ventral excitotoxic hippocampal lesions on conditioned freezing and spatial learning. Behav Neurosci 113:1189–1203
111. Rogers JL, Hunsaker MR, Kesner RP (2006) Effects of ventral and dorsal CA1 subregional lesions on trace fear conditioning. Neurobiol Learn Mem 86:72–81
112. Esclassan F, Coutureau E, Di Scala G, Marchand AR (2009) Differential contribution of dorsal and ventral hippocampus to trace and delay fear conditioning. Hippocampus 19:33–44
113. McHugh SB, Deacon RMJ, Rawlins JNP, Bannerman DM (2004) Amygdala and ventral hippocampal lesions contribute differentially to mechanisms of fear and anxiety. Behav Neurosci 118:63–78
114. Abrams JK, Johnson PL, Hay-Schmidt A, Mikkelsen J, Shekhar A, Lowry CA (2005) Serotonergic systems associated with arousal and vigilance behaviors following administration of anxiogenic drugs. Neuroscience 133:983–997
115. Lowry CA, Rodda JE, Lightman SL, Ingram CD (2000) Corticotropin-releasing factor increases in vitro firing rates of serotonergic neurons in the rat dorsal raphe nucleus: evidence for activation of a topographically organized mesolimbocortical serotonergic system. J Neurosci 20:7728–7736
116. Evans AK, Heerkens JLT, Lowry CA (2009) Acoustic stimulation in vivo and corticotropin-releasing factor in vitro increase tryptophan hydroxylase activity in the rat caudal dorsal raphe nucleus. Neurosci Lett 455:36–41
117. Singewald N, Sharp T (2000) Neuroanatomical targets of anxiogenic drugs in the hindbrain as revealed by Fos immunocytochemistry. Neuroscience 98:759–770
118. Hale MW, Stamper CE, Staub DR, Lowry CA (2010) Urocortin 2 increases c-Fos expression in serotonergic neurons projecting to the ventricular/periventricular system. Exp Neurol 224:271–281
119. Staub DR, Spiga F, Lowry CA (2005) Urocortin 2 increases c-Fos expression in topographically organized subpopulations of serotonergic neurons in the rat dorsal raphe nucleus. Brain Res 1044:176–189
120. Wright IK, Upton N, Marsden CA (1992) Effect of established and putative anxiolytics on extracellular 5-HT and 5-HIAA in the ventral hippocampus of rats during behaviour on the elevated X-maze. Psychopharmacology 109:338–346

121. Rex A, Voigt JP, Fink H (2005) Anxiety but not arousal increases 5-hydroxytryptamine release in the rat ventral hippocampus in vivo. Eur J Neurosci 22:1185–1189
122. Amat J, Matus-Amat P, Watkins LR, Maier SF (1998) Escapable and inescapable stress differentially and selectively and alter extracellular levels of 5-HT in the ventral hippocampus and dorsal periaqueductal gray of the rat. Brain Res 797:12–22
123. Okada M, Kawata Y, Murakami T, Wada K, Mizuno K, Kondo T, Kaneko S (1999) Differential effects of adenosine receptor subtypes on release and reuptake of hippocampal serotonin. Eur J Neurosci 11:1–9
124. Okada M, Nutt DJ, Murakami T, Zhu G, Kamata A, Kawata Y, Kaneko S (2001) Adenosine receptor subtypes modulate two major functional pathways for hippocampal serotonin release. J Neurosci 21:628–640
125. Okada M, Kawata Y, Kiryu K, Mizuno K, Wada K, Tasaki H, Kaneko S (1997) Effects of adenosine receptor subtypes on hippocampal extracellular serotonin level and serotonin reuptake activity. J Neurochem 69:2581–2588
126. Pascual O, Casper KB, Kubera C, Zhang J, Revilla-Sanchez R, Sul J-Y, Takano H, Moss SJ, McCarthy K, Haydon PG (2005) Astrocytic purinergic signaling coordinates synaptic networks. Science 310:113–116
127. Fontella FU, Bruno AN, Crema LM, Battastini AMO, Sarkis JJF, Netto CA, Dalmaz C (2004) Acute and chronic stress alter ecto-nucleotidase activities in synaptosomes from the rat hippocampus. Pharmacol Biochem Behav 78:341–347
128. Andrade R, Chaput Y (1991) 5-HT4-like receptors mediate the slow excitatory response to serotonin in the rat hippocampus. J Pharmacol Exp Ther 257:930–937
129. Andrade R, Nicoll RA (1987) Pharmacologically distinct actions of serotonin on single pyramidal neurones of the rat hippocampus recorded in vitro. J Physiol 394:99–124
130. Klemenhagen KC, Gordon JA, David DJ, Hen R, Gross CT (2006) Increased fear response to contextual cues in mice lacking the 5-HT1A receptor. Neuropsychopharmacology 31:101–111
131. Tsetsenis T, Ma XH, Lo Iacono L, Beck SG, Gross C (2007) Suppression of conditioning to ambiguous cues by pharmacogenetic inhibition of the dentate gyrus. Nat Neurosci 10:896–902
132. Nunes-de-Souza RL, Canto-de-Souza A, Rodgers RJ (2002) Effects of intra-hippocampal infusion of WAY-100635 on plus-maze behavior in mice. Influence of site of injection and prior test experience. Brain Res 927:87–96
133. File SE, Gonzalez LE (1996) Anxiolytic effects in the plus-maze of 5-HT1A-receptor ligands in dorsal raphé and ventral hippocampus. Pharmacol Biochem Behav 54:123–128
134. Hogg S, Andrews N, File SE (1994) Contrasting behavioural effects of 8-OHDPAT in the dorsal raphe nucleus and ventral hippocampus. Neuropharmacology 33:343–348
135. Cervo L, Mocaër E, Bertaglia A, Samanin R (2000) Roles of 5-HT1A receptors in the dorsal raphe and dorsal hippocampus in anxiety assessed by the behavioral effects of 8-OH-DPAT and S 15535 in a modified Geller-Seifter conflict model. Neuropharmacology 39:1037–1043
136. Felder CC, Kanterman RY, Ma AL, Axelrod J (1990) Serotonin stimulates phospholipase A2 and the release of arachidonic acid in hippocampal neurons by a type 2 serotonin receptor that is independent of inositolphospholipid hydrolysis. Proc Natl Acad Sci USA 87:2187–2191
137. Alves SH, Pinheiro G, Motta V, Landeira-Fernandez J, Cruz AP (2004) Anxiogenic effects in the rat elevated plus-maze of 5-HT(2C) agonists into ventral but not dorsal hippocampus. Behav Pharmacol 15:37–43
138. Masse F, Petit-Demouliere B, Dubois I, Hascöet M, Bourin M (2008) Anxiolytic-like effects of DOI microinjections into the hippocampus (but not the amygdala nor the PAG) in the mice four plates test. Behav Brain Res 188:291–297
139. Tecott LH, Logue SF, Wehner JM, Kauer JA (1998) Perturbed dentate gyrus function in serotonin 5-HT2C receptor mutant mice. Proc Natl Acad Sci USA 95:15026–15031
140. Kimura A, Stevenson PL, Carter RN, MacColl G, French KL, Simons JP, Al-Shawi R, Kelly V, Chapman KE, Holmes MC (2009) Overexpression of 5-HT2C receptors in forebrain leads to elevated anxiety and hypoactivity. Eur J Neurosci 30:299–306

141. Risold PY, Swanson LW (1997) Connections of the rat lateral septal nucleus. Brain Res Rev 24:115–195
142. Petrovich GD, Canteras NS, Swanson LW (2001) Combinatorial amygdalar inputs to hippocampal domains and hypothalamic behavior systems. Brain Res Rev 38:247–289
143. Risold PY (2004) The septal region. In: Paxinos G (ed) The rat nervous system, 3rd edn. Elsevier, New York, pp 605–632
144. Viana MB, Zangrossi H Jr, Onusic GM (2008) 5-HT1A receptors of the lateral septum regulate inhibitory avoidance but not escape behavior in rats. Pharmacol Biochem Behav 89:360–366
145. Pesold C, Treit D (1996) The neuroanatomical specificity of the anxiolytic effects of intra-septal infusions of midazolam. Brain Res 710:161–168
146. Cheeta S, Kenny PJ, File SE (2000) Hippocampal and septal injections of nicotine and 8-OH-DPAT distinguish among different animal tests of anxiety. Prog Neuropsychopharmacol Biol Psychiatry 24:1053–1067
147. Menard J, Treit D (1998) The septum and the hippocampus differentially mediate anxiolytic effects of R(+)-8-OH-DPAT. Behav Pharmacol 9:93–101
148. Hikosaka O, Sesack SR, Lecourtier L, Shepard PD (2008) Habenula: crossroad between the basal ganglia and the limbic system. J Neurosci 28:11825–11829
149. Matsumoto M, Hikosaka O (2007) Lateral habenula as a source of negative reward signals in dopamine neurons. Nature 447:1111–1115
150. Schultz W (1998) Predictive reward signal of dopamine neurons. J Neurophysiol 80:1–27
151. Matsumoto M, Hikosaka O (2009) Representation of negative motivational value in the primate lateral habenula. Nat Neurosci 12:77–84
152. Rausch LJ, Long CJ (1974) Habenular lesions and avoidance learning deficits in albine rats. Physiol Behav 2:352–356
153. Agetsuma M, Aizawa H, Aoki T, Nakayama R, Takahoko M, Goto M, Sassa T, Amo R, Shiraki T, Kawakami K, Hosoya T, Higashijima S-I, Okamoto H (2010) The habenula is crucial for experience-dependent modification of fear responses in zebrafish. Nat Neurosci 13:1354–1356
154. Lee A, Mathuru AS, Teh C, Kibat C, Korzh V, Penney TB, Jesuthasan S (2010) The habenula prevents helpless behavior in larval zebrafish. Curr Biol 20:2211–2216
155. Murphy CA, DiCamillo AM, Haun F, Murray M (1996) Lesion of the habenular efferent pathway produces anxiety and locomotor hyperactivity in rats: a comparison of the effects of neonatal and adult lesions. Behav Brain Res 81:43–52
156. Kim U (2009) Topographic commissural and descending projections of the habenula in the rat. J Comp Neurol 513:173–187
157. Wang RY, Aghajanian GK (1977) Physiological evidence for habenula as a major link between forebrain and midbrain raphe nucleus. Science 197:89–91
158. Ferraro G, Montalbano ME, Sardo P, La Grutta V (1996) Lateral habenular influence on dorsal raphe neurons. Brain Res Bull 41:47–52
159. Stern WC, Johnson A, Bronzino JD, Morgane PJ (1979) Effects of electrical stimulation of the lateral habenula on single-unit activity of raphe neurons. Exp Neurol 65:326–342
160. Neckers LM, Schwartz JP, Wyatt RJ, Speciale SG (1979) Substance P afferents from the habenula innervate the dorsal raphe nucleus. Exp Brain Res 37:619–623
161. Liu R, Ding Y, Aghajanian GK (2002) Neurokinins activate local glutamatergic inputs to serotonergic neurons of the dorsal raphe nucleus. Neuropsychopharmacology 27:329–340
162. Guiard BP, Guilloux J-P, Reperant C, Hunt SP, Toth M, Gardier AM (2007) Substance P neurokinin 1 receptor activation within the dorsal raphe nucleus controls serotonin release in the mouse frontal cortex. Mol Pharmacol 72:1411–1418
163. Grillon C (2008) Models and mechanisms of anxiety: evidence from startle studies. Psychopharmacology 199:421–437
164. Canteras NS (2002) The medial hypothalamic defensive system: hodological organization and functional implications. Pharmacol Biochem Behav 71:481–491

165. Brandão ML, Troncoso AC, Silva MAdS, Huston JP (2003) The relevance of neuronal substrates of defense in the midbrain tectum to anxiety and stress: empirical and conceptual considerations. Eur J Pharmacol 463:225–233
166. Holstege G (2009) The mesopontine rostromedial tegmental nucleus and the emotional motor system: role in basic survival behavior. J Comp Neurol 513:559–565
167. An X, Bandler R, Öngür D, Price JL (1998) Prefrontal cortical projections to longitudinal columns in the midbrain periaqueductal gray in macaque monkeys. J Comp Neurol 401:455–479
168. Bernard A, Bandler R (1998) Parallel circuits for emotional coping behaviour: new pieces in the puzzle. J Comp Neurol 401:429–436
169. Floyd NS, Price JL, Ferry AT, Keay KA, Bandler R (2000) Orbitomedial prefrontal cortical projections to distinct longitudinal columns of the periaqueductal gray in the rat. J Comp Neurol 422:556–578
170. Öngür D, An X, Price J (1998) Prefrontal cortical projections to the hypothalamus in macaque monkeys. J Comp Neurol 401:480–505
171. Keay KA, Bandler R (2004) Periaqueductal gray. In: Paxinos G (ed) The rat nervous system, 3rd edn. Academic, San Diego, pp 243–257
172. Bandler R, Shipley MT (1994) Columnar organization in the midbrain periaqueductal gray: modules for emotional expression? Trends Neurosci 17:379–389
173. Jhou TC, Fields HL, Baxter MG, Saper CB, Holland PC (2009) The rostromedial tegmental nucleus (RMTg), a GABAergic afferent to midbrain dopamine neurons, encodes aversive stimuli and inhibits motor responses. Neuron 61:786–800
174. Floyd NS, Price JL, Ferry AT, Keay KA, Bandler R (2001) Orbitomedial prefrontal cortical projections to hypothalamus in the rat. J Comp Neurol 432:307–328
175. Bielajew C, Jordan C, Ferme-Enright J, Shizgal P (1981) Refractory periods and anatomical linkage of the substrates for lateral hypothalamic and periaqueductal gray self-stimulation. Physiol Behav 27:95–104
176. Lowry CA (2002) Functional subsets of serotonergic neurones: implications for control of the hypothalamic-pituitary-adrenal axis. J Neuroendocrinol 14:911–923
177. Gray TS, Magnuson DJ (1987) Galanin-like immunoreactivity within amygdaloid and hypothalamic neurons that project to the midbrain central grey in rat. Neurosci Lett 83: 264–268
178. Gray TS, Magnuson DJ (1992) Peptide immunoreactive neurons in the amygdala and the bed nucleus of the stria terminalis project to the midbrain central gray in the rat. Peptides 13:451–460
179. Ljubic-Thibal V, Morin A, Diksic M, Hamel E (1999) Origin of the serotonergic innervation to the rat dorsolateral hypothalamus: retrograde transport of cholera toxin and upregulation of tryptophan hydroxylase mRNA expression following selective nerve terminals lesion. Synapse 32:177–186
180. Jasmin L, Burkey AR, Granato A, Ohara PT (2004) Rostral agranular insular cortex and pain areas of the central nervous system: a tract-tracing study in the rat. J Comp Neurol 468:425–440
181. Luiten PG, ter Horst GJ, Steffens AB (1987) The hypothalamus, intrinsic connections and outflow pathways to the endocrine system in relation to the control of feeding and metabolism. Prog Neurobiol 28:1–54
182. Rompré P–P, Miliaressis E (1985) Pontine and mesencephalic substrates of self-stimulation. Brain Res 359:246–259
183. Saper CB, Swanson LW, Cowan WM (1979) An autoradiographic study of the efferent connections of the lateral hypothalamic area in the rat. J Comp Neurol 183:689–706
184. Stezhka VV, Lovick TA (1997) Projections from dorsal raphe nucleus to the periaqueductal grey matter: studies in slices of rat midbrain maintained in vitro. Neurosci Lett 230:57–60
185. Hunsperger RW (1956) Affektreaktionen auf elektrische Reizung im Hirnstamm der Katze. Helvetica Physiologica et Pharmacologica Acta 14:70–92

186. Fernandez de Molina A, Hunsperger RW (1962) Organization of the subcortical system governing defense and flight reactions in the cat. J Physiol 160:200–213
187. Brown SM, Peet E, Manuck SB, Williamson DE, Dahl RE, Ferrell RE, Hariri AR (2005) A regulatory variant of the human tryptophan hydroxylase-2 gene biases amygdala reactivity. Mol Psychiatry 10:884–888
188. Adamec R (2000) Evidence that long-lasting potentiation of amygdala efferents in the right hemisphere underlies pharmacological stressor (FG-7142) induced lasting increases in anxiety-like behaviour: role of GABA tone in initiation of brain and behavioural changes. J Psychopharmacol 14:323–339
189. Pavlova IV (2006) Linkage of neuron spike activity in the right and left amygdala in food motivation and emotional tension. Neurosci Behav Physiol 36:217–225
190. Adamec R, Blundell J, Burton P (2003) Phosphorylated cyclic AMP response element binding protein expression induced in the periaqueductal gray by predator stress: its relationship to the stress experience, behavior and limbic neural plasticity. Prog Neuropsychopharmacol Biol Psychiatry 27:1243–1267
191. Adamec R, Berton O, Razek WA (2009) Viral vector induction of CREB expression in the periaqueductal gray induces a predator stress-like pattern of changes in pCREB expression, neuroplasticity, and anxiety in rodents. Neural Plasticity 2009: Article ID 904568
192. Adamec R (1998) Evidence that NMDA dependent limbic neural plasticity in the right hemisphere mediates pharmacological stressor (FG-7142) induced lasting increases in anxiety-like behavior sutdy 1—Role of NMDA receptors in efferent transmission from the cat amygdala. J Psychopharmacol 12:122–128
193. Adamec R (1998) Evidence that NMDA dependent limbic neural plasticity in the right hemisphere mediates pharmacological stressor (FG-7142) induced lasting increases in anxiety-like behavior study 3—The effects on amygdala efferent physiology of block of NMDA receptors just prior to behavioral change. J Psychopharmacol 12:227–238
194. Paré D, Smith Y (1993) The intercalated cell masses project to the central and medial nuclei of the amygdala in cats. Neuroscience 57:1077–1090
195. Royer S, Martina M, Paré D (1999) An inhibitory interface gates impulse traffic between the input and output stations of the amygdala. J Neurosci 19:10575–10583
196. Gasser PJ, Orchinik M, Raju I, Lowry CA (2009) Distribution of organic cation transporter 3, a corticosterone-sensitive monoamine transporter, in the rat brain. J Comp Neurol 512:529–555
197. Quirk GJ, Likhtik E, Pelletier JG, Paré D (2003) Stimulation of medial prefrontal cortex decreases the responsiveness of central amygdala output neurons. J Neurosci 23:8800–8807
198. Gozzi A, Jain A, Giovanelli A, Bertollini C, Crestan V, Schwarz AJ, Tsetsenis T, Ragozzino D, Gross CT, Bifone A (2010) A neural switch for active and passive fear. Neuron 67:656–666
199. Paxinos G, Watson C (1998) The rat brain in stereotaxic coordinates, 4th edn. Academic, San Diego
200. Cassell MD, Freedman LJ, Shi C (1999) The intrinsic organization of the central extended amygdala. Ann N Y Acad Sci 877:217–241
201. Spannuth BM, Hale MW, Evans AK, Lukkes JL, Campeau S, Lowry CA (2011) Investigation of a central nucleus of the amygdala/dorsal raphe nucleus serotonergic circuit implicated in fear-potentiated startle. Neuroscience 179:104–119
202. Huber D, Veinante P, Stoop R (2005) Vasopressin and oxytocin excite distinct neuronal populations in the central amygdala. Science 308:245–248
203. Ciocchi S, Herry C, Grenier F, Wolff SBE, Letzkus JJ, Vlachos I, Ehrlich I, Sprengel R, Deisseroth K, Stadler MB, Müller C, Lüthi A (2010) Encoding of conditioned fear in central amygdala inhibitory circuits. Nature 468:277–282
204. Haubensak W, Kunwar PS, Cai H, Ciocchi S, Wall NR, Ponnusamy R, Biag J, Dong H-W, Deisseroth K, Callaway EM, Fanselow MS, Lüthi A, Anderson DJ (2011) Genetic dissection of an amygdala microcircuit that gates conditioned fear. Nature 468:270–276
205. Scicli AP, Petrovich GD, Swanson LW, Thompson RF (2004) Contextual fear conditioning is associated with lateralized expression of the immediate early gene c-fos in the central and basolateral amygdalar nuclei. Behav Neurosci 118:5–14

206. Malkani S, Rosen JB (2000) Differential expression of EGR-1 mRNA in the amygdala following diazepam in contextual fear conditioning. Brain Res 860:53–63
207. Beck CHM, Fibiger HC (1995) Conditioned fear-induced changes in behavior and in the expression of the immediate early gene c-fos: with and without diazepam pretreatment. J Neurosci 15:709–720
208. Medeiros MA, Reis LC, Mello LE (2005) Stress-induced c-Fos expression is differentially modulated by dexamethasone, diazepam and imipramine. Neuropsychopharmacology 30:1246–1256
209. Day HEW, Kryskow EM, Nyhuis TJ, Herlihy L, Campeau S (2008) Conditioned fear inhibits c-fos mRNA in the central extended amygdala. Brain Res 1229:137–146
210. Duvarci S, Popa D, Paré D (2011) Central amygdala activity during fear conditioning. J Neurosci 31:289–294
211. Tkacs NC, Li J, Strack AM (1997) Central amygdala Fos expression during hypotensive or febrile, nonhypotensive endotoxemia in conscious rats. J Comp Neurol 379:592–602
212. Torres G, Horowitz JM, Laflamme N, Rivest S (1998) Fluoxetine induces the transcription of genes encoding c-fos, corticotropin-releasing factor and its type 1 receptor in rat brain. Neuroscience 87:463–477
213. Beck CHM (1995) Acute treatment with antidepressant drugs selectively increases the expression of c-fos in the rat brain. J Psychiatry Neurosci 20:25–32
214. Slattery DA, Morrow JA, Hudson AL, Hill DR, Nutt DJ, Henry B (2005) Comparison of alterations in c-fos and Egr-1 (zif268) expression throughout the rat brain following acute administration of different classes of antidepressant compounds. Neuropsychopharmacology 30:1278–1287
215. Holahan MR, White NM (2004) Amygdala c-Fos induction corresponds to unconditioned and conditioned aversive stimuli but not to freezing. Behav Brain Res 152:109–120
216. Heisler LK, Zhou L, Bajwa P, Hsu J, Tecott LH (2007) Serotonin 5-HT2C receptors regulate anxiety-like behavior. Genes, Brain and Behavior 6:491–496
217. Touzani K, Taghzouti K, Velley L (1997) Increase of the aversive value of taste stimuli following ibotenic acid lesion of the central amygdaloid nucleus in the rat. Behav Brain Res 88:133–142
218. Koo JW, Han J-S, Kim JJ (2004) Selective neurotoxic lesions of basolateral and central nuclei of the amygdala produce differential effects on fear conditioning. J Neurosci 24:7654–7662
219. Campeau S, Davis M (1995) Involvement of the central nucleus and basolateral complex of the amygdala in fear conditioning measured with fear-potentiated startle in rats trained concurrently with auditory and visual conditioned stimuli. J Neurosci 15:2301–2311
220. Kim M, Campeau S, Falls WA, Davis M (1993) Infusion of the non-NMDA receptor antagonist CNQX into the amygdala blocks the expression of fear-potentiated startle. Behav Neural Biology 59:5–8
221. Rosen JB, Hitchcock JM, Sananes CB, Miserendino MJ, Davis M (1991) A direct projection from the central nucleus of the amygdala to the acoustic startle pathway: anterograde and retrograde tracing studies. Behav Neurosci 105:817–825
222. Rizvi TA, Ennis M, Behbehani MM, Shipley MT (1991) Connections between the central nucleus of the amygdala and the midbrain periaqueductal gray: topography and reciprocity. J Comp Neurol 303:121–131
223. Viviani D, Charlet A, van den Burg E, Robinet C, Hurni N, Abatis M, Magara F, Stoop R (2011) Oxytocin selectively gates fear responses through distinct outputs from the central amygdala. Science 333:104–107
224. Kalin NH, Shelton SE, Davidson RJ (2004) The role of the central nucleus of the amygdala in mediating fear and anxiety in the primate. J Neurosci 24:5506–5515
225. Moreira CM, Masson S, Carvalho MC, Brandão ML (2007) Exploratory behaviour of rats in the elevated plus-maze is differentially sensitive to inactivation of the basolateral and central amygdaloid nuclei. Brain Res Bull 71:466–474

226. Asan E, Yilmazer-Hanke DM, Eliava M, Hantsch M, Lesch K-P, Schmitt A (2005) The corticotropin-releasing factor (CRF)-system and monoaminergic afferents in the central amygdala: investigations in different mouse strains and comparison with the rat. Neuroscience 131:953–967
227. Hale MW, Lowry CA (2011) Functional topography of midbrain and pontine serotonergic systems: implications for synaptic regulation of serotonergic circuits. Psychopharmacology 213:243–264
228. Lowry CA, Jonhson PL, Hay-Schmidt A, Mikkelsen J, Shekhar A (2005) Modulation of anxiety circuits by serotonergic systems. Stress 8:233–246
229. Bard PA (1928) A diencephalic mechanism for the expression of rage with special reference to the sympathetic nervous system. Am J Physiol 84:490–510
230. Lipp HP, Hunsperger RW (1978) Threat, attack and flight elicited by electrical stimulation of ventromedial hypothalamus of marmoset monkeys Callitrhix jacchus. Brain Behav Evol 15:260–293
231. Azevedo AD, Hilton SM, Timms RJ (1980) The defense reaction elicited by midbrain and hypothalamic stimulation in the rabbit. J Physiol 301:56–57
232. Brutus M, Shaikh MB, Siegel A (1985) Differential control of hypothalamically elicited flight behavior by the midbrain periaqueductal gray in the cat. Behav Brain Res 17:235–244
233. Fuchs SAG, Edinger HM, Siegel A (1985) The organization of the hypothalamic pathways mediating affective defensive behavior in the cat. Brain Res 330:77–92
234. Yardley CP, Hilton SM (1986) The hypothalamic and brainstem areas from which the cardiovascular and behavioural components of the defense reaction are elicited in the rat. J Auton Nerv Syst 15:227–244
235. Lammers JHCM, Kruk MR, Meelis W, van der Poel AM (1988) Hypothalamic substrates for brain stimulation-induced patterns of locomotion and escape jumps in the rat. Brain Res 449:294–310
236. Silveira MCL, Graeff FG (1992) Defense reaction elicited by microinjection of kainic acid into the medial hypothalamus of the rat: antagonism by GABAA receptor agonist. Behav Neural Biol 57:226–232
237. Di Scala G, Schmitt P, Karli P (1984) Flight induced by infusion of bicuculline methiodide into periventricular structures. Brain Res 309:199–208
238. Brandão ML, Di Scala G, Bouchet MJ, Schmitt P (1986) Escape behavior produced by blockade of glutamic acid decarboxylase (GAD) in mesencephalic central gray or medial hypothalamus. Pharmacol Biochem Behav 24:497–501
239. Milani H, Graeff FG (1987) GABA-benzodiazepine modulation of aversion in the medial hypothalamus of the rat. Pharmacol Biochem Behav 28:21–27
240. Schmitt P, Di Scala G, Brandão ML, Karli P (1985) Behavioral effects of microinjections of SR 95103, a new GABA-A antagonist, into the medial hypothalamus or the mesencephalic central gray. Eur J Pharmacol 117:149–158
241. Canteras NS, Chiavegatto S, Ribeiro do Valle LE, Swanson LW (1997) Severe reduction of defensive behavior to a predator by discrete hypothalamic chemical lesions. Brain Res Bull 44:297–305
242. Dielenberg RA, Hunt GE, McGregor IS (2001) 'When a rat smells a cat': the distribution of fos immunoreactivity in rat brain following exposure to a predatory odor. Neuroscience 104:1085–1097
243. Vianna DM, Borelli KG, Ferreira-Netto C, Macedo CE, Brandão ML (2003) Fos-like immunoreactive neurons following electrical stimulation of the dorsal periaqueductal gray at freezing and escape thresholds. Brain Res Bull 62:179–189
244. Singewald N (2007) Altered brain activity processing in high-anxiety rodents revealed by challenge paradigms and functional mapping. Neurosci Biobehav Rev 31:18–40
245. Kanarik M, Alttoa A, Matrov D, Kõiv K, Sharp T, Panksepp J, Harro J (2011) Brain responses to chronic social defeat stress: effects on regional oxidative metabolism as a function of a hedonic trait, and gene expression in suceptible and resilient rats. Eur Neuropsychopharmacol 21:92–107

246. Staples LG, Hunt GE, Cornish JL, McGregor IS (2005) Neural activation during cat odor-induced conditioned fear and 'trial 2' fear in rats. Neurosci Biobehav Rev 29:1265–1277
247. Cezario AF, Ribeiro-Barbosa ER, Baldo MVC, Canteras NS (2008) Hypothalamic sites responding to predator threats—The role of the dorsal premammillary nucleus in unconditioned and conditioned antipredatory defensive behavior. Eur J Neurosci 28: 1003–1015
248. DiMicco JA, Samuels BC, Zaretskaia MV, Zaretsky DV (2002) The dorsomedial hypothalamus and the response to stress. Part renaissance, part revolution. Pharmacol Biochem Behav 71:469–480
249. Singewald N, Salchner P, Sharp T (2003) Induction of c-fos expression in specific areas of the fear circuitry in rat forebrain by anxiogenic drugs. Biol Psychiatry 53:275–283
250. Vialou V, Balasse L, Dumas S, Giros B, Gautron S (2007) Neurochemical characterization of pathways expressing plasma membrane monoamine transporter in the rat brain. Neuroscience 144:616–622
251. Lowry CA, Plant A, Shanks N, Ingram CD, Lightman SL (2003) Anatomical and functional evidence for a stress-responsive, monoamine-accumulating area in the dorsomedial hypothalamus of adult rat brain. Horm Behav 43:254–262
252. Feng N, Mo B, Johnson PL, Orchinik M, Lowry CA, Renner KJ (2005) Local inhibition of organic cation transporters increases extracellular serotonin in the medial hypothalamus. Brain Res 1063:69–76
253. Feng N, Lowry CA, Lukkes JL, Orchinik M, Forster GL, Renner KJ (2010) Organic cation transporter inhibition increases medial hypothalamic serotonin under basal conditions and during mild restraint. Brain Res 1326:105–113
254. Lowry CA, Burke KA, Renner KJ, Moore FL, Orchinik M (2001) Rapid changes in monoamine levels following administration of corticotropin-releasing factor or corticosterone are localized in the dorsomedial hypothalamus. Horm Behav 39:195–205
255. Clements S, Moore FL, Schreck CB (2003) Evidence that acute serotonergic activation potentiates the locomotor-stimulating effects of corticotropin-releasing hormone in juvenile Chinook salmon (Oncorhynchus tshawytscha). Horm Behav 43:214–221
256. Shekhar A, Keim SR, Simon JR, McBride WJ (1996) Dorsomedial hypothalamic GABA dysfunction produces physiological arousal following sodium lactate infusions. Pharmacol Biochem Behav 55:249–256
257. Molosh AI, Johnson PL, Fitz SD, DiMicco JA, Herman JP, Shekhar A (2010) Changes in central sodium and not osmolarity or lactate induce panic-like responses in a model of panic disorder. Neuropsychopharmacology 35:1333–1347
258. Johnson PL, Lowry CA, Truitt W, Shekhar A (2008) Disruption of GABAergic tone in the dorsomedial hypothalamus attenuates responses in a subset of serotonergic neurons in the dorsal raphe nucleus following lactate-induced panic. J Psychopharmacol 22:642–652
259. Morin SM, Stotz-Potter EH, DiMicco JA (2001) Injection of muscimol into dorsomedial hypothalamus and stress-induced Fos expression in paraventricular nucleus. Am J Physiol 280:R1276–R1284
260. Stotz-Potter EH, Morin SM, DiMicco JA (1996) Effect of microinjection of muscimol into dorsomedial or paraventricular hypothalamic nucleus on stress-induced neuroendocrine and cardiovascular changes in rats. Brain Res 1996:742
261. Stotz-Potter EH, Willis LR, DiMicco JA (1996) Muscimol acts in dorsomedial and not paraventricular hypothalamic nucleus to suppress cardiovascular effects of stress. J Neurosci 16:1173–1179
262. Engelmann M, Ebner K, Landgraf R, Holsboer CT, Wotjak CT (1999) Emotional stress triggers intrahypothalamic but not peripheral release of oxytocin in male rats. J Neuroendocrinol 11:867–872
263. Wotjak CT, Ganster J, Kohl G, Holsboer R, Landgraf R, Engelmann M (1998) Dissociated central and peripheral release of vasopressin, but not oxytocin, in response to repeated swim stress: new insights into the secretory capacities of peptidergic neurons. Neuroscience 85: 1209–1222

264. Wotjak CT, Kubota M, Liebsch G, Montkowski A, Holsboer F, Neumann I, Landgraf R (1996) Release of vasopressin within the rat paraventricular nucleus in response to emotional stress: a novel mechanism of regulating adrenocorticotropic hormone secretion? J Neurosci 16:7725–7732
265. Makara GB, Antoni FA, Stark E (1982) Electrical stimulation in the rat of the supraoptic nucleus: failure to alter plasma corticosterone after surgical lesioning of the paraventricular nucleus. Neurosci Lett 30:269–273
266. Engelmann M, Landgraf R, Wotjak CT (2004) The hypothalamic-neurohypophysial system regulates the hypothalamic-pituitary-adrenal axis under stress: an old concept revisited. Front Neuroendocrinol 25:132–149
267. Lee S, Rivier C (1998) Interaction between corticotropin-releasing factor and nitric oxide in mediating the response of the rat hypothalamus to immune and non-immune stimuli. Mol Brain Res 57:54–62
268. Di S, Malcher-Lopes R, Halmos KC, Tasker JG (2003) Nongenomic glucocorticoid inhibition via endocannabinoid release in the hypothalamus: a fast feedback mechanism. J Neurosci 23:4850–4857
269. Carrasco GA, Van de Kar LD (2003) Neuroendocrine pharmacology of stress. Eur J Pharmacol 463:235–272
270. Holmes A, Li Q, Murphy DL, Gold E, Crawley JN (2003) Abnormal anxiety related behavior in serotonin transporter null mutant mice: the influence of genetic background. Genes Brain Behav 2:365–380
271. Holmes A, Yang RJ, Lesch K-P, Crawley JN, Murphy DL (2003) Mice lacking the serotonin transporter exhibit 5-HT1A receptor-mediated abnormalities in tests of anxiety-like behavior. Neuropsychopharmacology 28:2077–2088
272. Li Q, Wichems C, Heils A, Van de Kar LD, Lesch K-P, Murphy DL (1999) Reduction of 5-hydroxytryptamine (5-HT)(1A)-mediated temperature and neuroendocrine responses and 5-HT(1A) binding sites in 5-HT transporter knockout mice. J Pharmacol Exp Ther 291: 999–1007
273. Li Q, Holmes A, Ma L, Van de Kar LD, Garcia F, Murphy DL (2004) Medial hypothalamic 5-hydroxytryptamine (5-HT)1A receptors regulate neuroendocrine responses to stress and exploratory locomotor activity: application of recombinant adenovirus containing 5-HT1A sequences. J Neurosci 24:10868–10877
274. Van de Kar LD, Javed A, Zhang Y, Serres F, Raap DK, Gray TS (2001) 5-HT2A receptors stimulate ACTH, corticosterone, oxytocin, renin, and prolactin release and activate hypothalamic CRF and oxytocin-expressing cells. J Neurosci 21:3572–3579
275. Kaufling J, Veinante P, Pawlowski SA, Freund-Mercier M-J, Barrot M (2009) Afferents to the GABAergic tail of the ventral tegmental area in the rat. J Comp Neurol 513:597–621
276. Herbert H, Saper CB (1992) Organization of medullary adrenergic and noradrenergic projections to the periaqueductal gray matter in the rat. J Comp Neurol 315:34–52
277. Nagy JI, Patel D, Ochalski PAY, Stelmack GL (1998) Connexin30 in rodent, cat and human brain: selective expression in gray matter astrocytes, co-localization with connexin43 at gap junctions, and late developmental appearance. Neuroscience 88:447–468
278. Fabbri A, Fraioli F, Pert CB, Pert A (1985) Calcitonin receptors in the rat mesencephalon mediate its analgesic actions: autoradiographic and behavioral analyses. Brain Res 343: 205–215
279. Keay KA, Clement CI, Depaulis A, Bandler R (2001) Different representations of inescapable noxious stimuli in the periaqueductal gray and upper cervical spinal cord of freely moving rats. Neurosci Lett 313:17–20
280. Silveira MCL, Zangrossi H Jr, Viana MB, Silveira R, Graeff FG (2001) Differential expression of Fos protein in the rat brain induced by performance of avoidance or escape in the elevated T-maze. Behav Brain Res 126:13–21
281. Adamec R, Toth M, Haller J, Halasz J, Blundell J (2010) Activation patterns of cells in selected brain stem nuclei of more or less stress responsive rats in two animal models of PTSD—Predator exposure and submersion stress. Neuropharmacology 62:725–736

282. Berton O, Covington HE 3rd, Ebner L, Tsankova NM, Carle TL, Ulery P, Bhonsle A, Barrot M, Krishnan V, Singewald GM, Singewald N, Birnbaum S, Neve RL, Nestler EJ (2007) Induction of FosB in the periaqueductal gray by stress promotes active coping responses. Neuron 55:289–300
283. Vieira EB, Menescal-de-Oliveira L, Leite-Panissi CRA (2011) Functional mapping of the periaqueductal gray matter involved in organizing tonic immobility behavior in guinea pigs. Behav Brain Res 216:94–99
284. Carrive P, Leung P, Harris J, Paxinos G (1997) Conditioned fear to context is associated with increased Fos expression in the caudal ventrolateral region of the midbrain periaqueductal gray. Neuroscience 78:165–177
285. Hacker J, Pedersen NP, Chieng BCH, Keay KA, Christie MJ (2006) Enhanced Fos expression in glutamic acid decarboxylase immunoreactive neurons of the mouse periaqueductal gray during opioid withdrawal. Neuroscience 137:1389–1396
286. McGregor IS, Hargreaves GA, Apfelbach R, Hunt GE (2004) Neural correlates of cat odor-induced anxiety in rats: region-specific effects of the benzodiazepine midazolam. J Neurosci 24:4134–4144
287. Rosen JB, Adamec RE, Thompson BL (2005) Expression of egr-1 (zif268) mRNA in select fear-related brain regions following exposure to a predator. Behav Brain Res 162:279–288
288. Staples LG, McGregor IS, Hunt GE (2009) Long-lasting FosB/FosB immunoreactivity in the rat brain after repeated cat odor exposure. Neurosci Lett 462:157–161
289. Coimbra NC, Brandão ML (1993) GABAergic nigro-collicular pathways modulate the defensive behavior elicited by midbrain tectum stimulation. Behav Brain Res 59:131–139
290. Schenberg LC, Costa MB, Borges PCL, Castro FS (1990) Logistic analysis of the defense reaction induced by electrical stimulation of the rat mesencephalic tectum. Neurosci Biobehav Rev 14:473–479
291. Schenberg LC, Bittencourt AS, Sudré EC, Vargas LC (2001) Modeling panic attacks. Neurosci Biobehav Rev 25:647–659
292. Depaulis A, Keay KA, Bandler R (1994) Quiescence and hypo-reactivity evoked by activation of cell bodies in the ventrolateral midbrain periaqueductal gray of the rat. Exp Brain Res 99:75–83
293. Zhang SP, Bandler R, Carrive P (1990) Flight and immobility evoked by excitatory amino acid microinjection within distinct parts of the subtentorial midbrain periaqueductal gray of the cat. Brain Res 520:73–82
294. Monassi CR, Leite-Panissi CRA, Menescal-de-Oliveira L (1999) Ventrolateral periaqueductal gray matter and control of tonic immobility. Brain Res Bull 50:201–208
295. Vianna DML, Graeff FG, Brandão ML, Landeira-Fernandez J (2001) Defensive freezing evoked by electrical stimulation of the periaqueductal gray: comparison between dorsolateral and ventrolateral regions. NeuroReport 12:4109–4112
296. Fanselow MS, DeCola JP, De Oca BM, Landeira-Fernandez J (1995) Ventral and dorsolateral regions of the midbrain periaqueductal gray (PAG) control different stages of defensive behavior: dorsolateral PAG lesions enhance the defensive freezing produced by massed and immediate shock. Aggressive Behav 21:63–77
297. Magierek V, Ramos PL, Silveira-Filho NG, Nogueira RL, Landeira-Fernandez J (2003) Context fear conditioning inhibits panic-like behavior elicited by electrical stimulation of dorsal periaqueductal gray. NeuroReport 14:1641–1644
298. Vianna DML, Graeff FG, Landeira-Fernandez J, Brandão ML (2001) Lesions of the ventral periaqueductal gray reduces conditioned fear but does not change freezing induced by stimulation of the dorsal periaqueductal gray. Learn Mem 8:164–169
299. Griffiths JL, Lovick TA (2002) Co-localization of 5-HT2A-receptor and GABA immunoreactivity in neurones in the periaqueductal grey matter of the rat. Neurosci Lett 326:151–154
300. Lovick TA, Paul NL (1999) Co-localization of GABA with nicotinamide adenine dinucleotide phosphate-dependent diaphorase in neurones in the dorsolateral periaqueductal grey of the rat. Neurosci Lett 272:167–170

301. Brandão ML, Aguiar JC, Graeff FG (1982) GABA mediation of the antiaversive action of minor tranquilizers. Pharmacol Biochem Behav 16:397–402
302. Motta V, Brandão ML (1993) Aversive and antiaversive effects of morphine in the dorsal periaqueductal gray of rats submitted to the elevated plus-maze test. Pharmacol Biochem Behav 44:119–125
303. Russo AS, Guimarães FS, Aguiar JC, Graeff FG (1993) Role of benzodiazepine receptors located in the dorsal periaqueductal grey of rats in anxiety. Psychopharmacology 110:198–202
304. Reimer AE, Oliveira AR, Brandão ML (2009) Involvement of GABAergic mechanisms of the dorsal periaqueductal gray and inferior colliculus on unconditioned fear. Psychol Neurosci 2:51–58
305. Oliveira LC, Broiz AC, Macedo CE, Landeira-Fernandez J, Brandão ML (2006) 5-HT2 receptor mechanisms of the dorsal periaqueductal gray in the conditioned and unconditioned fear in rats. Psychopharmacology 191:253–262
306. Zanovelli JM, Nogueira RL, Zangrossi H Jr (2003) Serotonin in the dorsal periaqueductal gray modulates inhibitory avoidance and one-way escape behaviors in the elevated T-maze. Eur J Pharmacol 473:153–161
307. VdP Soares, Zangrossi H Jr (2004) Involvement of 5-HT1A and 5-HT2 receptors of the dorsal periaqueductal gray in the regulation of the defensive behaviors generated by the elevated T-maze. Brain Res Bull 64:181–188
308. Castilho VM, Brandão ML (2001) Conditioned antinociception and freezing using electrical stimulation of the dorsal periaqueductal gray or inferior colliculus as unconditioned stimulus are differentially regulated by 5-HT2A receptors in rats. Psychopharmacology 155:154–162
309. Monassi CR, Menescal-de-Oliveira L (2004) Serotonin 5-HT2 and 5-HT1A receptors in the periaqueductal gray matter differentially modulate tonic immobility in guinea pig. Brain Res 1009:169–180
310. Gomes KS, Nunes-de-Souza RL (2009) Implication of the 5-HT2A and 5-HT2C (but not 5-HT1A) receptors located within the periaqueductal gray in the elevated plus-maze test-retest paradigm in mice. Prog Neuropsychopharmacol Biol Psychiatry 33:1261–1269
311. Nunes-de-Souza V, Nunes-de-Souza RL, Rodgers RJ, Canto-de-Souza A (2008) 5-HT2 receptor activation in the midbrain periaqueductal gray (PAG) reduces anxiety-like behaviour in mice. Behav Brain Res 187:72–79
312. Moreira FA, Guimarães FS (2004) Benzodiazepine receptor and serotonin 2A receptor modulate the aversive-like effects of nitric oxide in the dorsolateral periaqueductal gray of rats. Psychopharmacology 176:362–368
313. Brandão ML, Zanovelli JM, Ruiz-Martinez RC, Oliveira LC, Landeira-Fernandez J (2008) Different patterns of freezing behavior organized in the periaqueductal gray of rats: association with different types of anxiety. Behav Brain Res 188:1–13
314. Xing J, Lu J, Li J (2008) Purinergic P2X receptors presynaptically increase glutamatergic synaptic transmission in dorsolateral periaqueductal gray. Brain Res 1208:46–55
315. Brandão ML, Lopez-Garcia JA, Roberts MHT (1991) Electrophysiological evidence for excitatory 5-HT2 and depressant 5-HT1A receptors on neurones of the rat midbrain tectum. Brain Res 556:259–266
316. Beckett SRG, Marsden CA, Marshall PW (1992) Attenuation of chemically induced defence responses by 5-HT1 receptor agonists administered into the periaqueductal gray. Psychopharmacology 108:110–114
317. Broiz AC, Oliveira LC, Brandão ML (2008) Regulation of conditioned and unconditioned fear in rats by 5-HT1A receptors in the dorsal periaqueductal gray. Pharmacol Biochem Behav 89:76–84
318. Lovick TA, Parry DM, Stezhka VV, Lumb BM (1999) Serotonergic transmission in the periaqueductal gray matter in relation to aversive behavior: morphological evidence for direct modulatory effects on identified output neurons. Neuroscience 95:763–772

319. Miguel TLB, Pobbe RLH, Spiaci A Jr, Zangrossi H Jr (2010) Dorsal raphe nucleus regulation of a panic-like defensive behavior evoked by chemical stimulation of the rat dorsal periaqueductal gray matter. Behav Brain Res 213:195–200
320. Fields HL, Bry J, Hentall I, Zorman G (1983) The activity of neurons in the rostral medulla of the rat during withdrawal from noxious heat. J Neurosci 3:2545–2552
321. Cheng ZF, Fields HL, Heinricher MM (1986) Morphine microinjected into the periaqueductal gray has differential effects on 3 classes of medullary neurons. Brain Res 375:57–65
322. Heinricher MM, Morgan MM, Tortorici V, Fields HL (1994) Disinhibition of off-cells and antinociception produced by an opioid action within the rostral ventromedial medulla. Neuroscience 63:279–288
323. McGaraughty S, Heinricher MM (2002) Microinjection of morphine into various amygdaloid nuclei differentially affects nociceptive responsiveness and RVM neuronal activity. Pain 96:153–162
324. Barbaro NM, Heinricher MM, Fields HL (1989) Putative nociceptive modulating neurons in the rostral ventromedial medulla of the rat: firing of on- and off-cells is related to nociceptive responsiveness. Somatosens Morot Res 6:427–439
325. Heinricher MM, Morgan MM, Fields HL (1992) Direct and indirect actions of morphine on medullary neurons that modulate nociception. Neuroscience 48:533–543
326. Sewards TV, Sewards MA (2002) The medial pain system: neural representations of the motivational aspect of pain. Brain Res Bull 59:163–180
327. Behbehani MM (1995) Functional characteristics of the midbrain periaqueductal gray. Prog Neurobiol 46:575–605
328. Nichols DS, Thron BE, Berntson GG (1989) Opiate and serotonergic mechanisms of stimulation-produced analgesia within the periaqueductal gray. Brain Res Bull 22:717–724
329. Schul R, Frenk H (1991) The role of serotonin in analgesia elicited by morphine in the periaqueductal gray matter (PAG). Brain Res 553:353–357
330. de Luca-Vinhas MCZ, Brandão ML, Motta VA, Landeira-Fernandez J (2003) Antinociception induced by stimulation of ventrolateral periaqueductal at the freezing threshold is regulated by opioid and 5-HT2A receptors as assessed by the tail-flick and formalin tests. Pharmacol Biochem Behav 75:459–466
331. Oliveira MA, Prado WA (2001) Role of PAG in the antinociception evoked from the medial or central amygdala in rats. Brain Res Bull 54:55–63
332. VdP Soares, Zangrossi H Jr (2009) Stimulation of 5-HT1A or 5-HT2A receptors in the ventrolateral periaqueductal gray causes anxiolytic-, but not panicolytic-like effect in rats. Behav Brain Res 197:178–185
333. De Luca-Vinhas MCZ, Macedo CE, Brandão ML (2006) Pharmacological assessment of the freezing, antinociception, and exploratory behavior organized in the ventrolateral periaqueductal gray. Pain 121:94–104
334. Aston-Jones G, Foote SL, Segal M (1985) Impulse conduction properties of noradrenergic locus coeruleus axons projecting to monkey cerebrocortex. Neuroscience 15:765–777
335. Lewis DA, Campbell MJ, Foote SL, Goldstein M, Morrison JH (1987) The distribution of tyrosine hydroxylase-immunoreactive fibers in primate neocortex is widespread but regionally specific. J Neurosci 7:279–290
336. Elam M, Yao T, Svensson TH, Thoren P (1984) Regulation of locus coeruleus neurons and splanchnic, sympathetic nerves by cardiovascular afferents. Brain Res 290:281–297
337. Berridge CW, Waterhouse BD (2003) The locus coeruleus-noradrenergic system: modulation of behavioral state and state-dependent cognitive processes. Brain Res Rev 42:33–84
338. Alvarez-Maubecin V, Garcia-Hernandez F, Williams JT, Van Bockastaele EJ (2000) Functional coupling between neurons and glia. J Neurosci 20:4091–4098
339. Usher M, Cohen JD, Rajowski J, Kubiak P, Aston-Jones G (1999) The role of locus coeruleus in the regulation of cognitive performance. Science 283:549–554

340. Van Bockastaele EJ, Bajic D, Proudfit H, Valentino RJ (2001) Topographic architecture of stress-related pathways targeting the noradrenergic locus coeruleus. Physiol Behav 73:273–283
341. Aston-Jones G, Ennis M, Pieribone VA, Nickell WT, Shipley MT (1986) The brain nucleus locus coeruleus: restricted afferent control of a broad efferent network. Science 234:734–737
342. Van Bockastaele EJ, Peoples J, Valentino RJ (1999) Anatomic basis for differential regulation of the rostrolateral peri-locus coeruleus region by limbic afferents. Biol Psychiatry 46:1352–1363
343. Toth ZE, Gallatz K, Fodor M, Palkovits M (1999) Decussations of the descending paraventricular pathways to the brainstem and spinal cord autonomic centers. J Comp Neurol 414:255–266
344. Van Bockastaele EJ, Colago EE, Valentino RJ (1996) Corticotropin-releasing factor-containing axon terminals synapse onto catecholamine dendrites and may presynaptically modulate other afferents in the rostral pole of the nucleus locus coeruleus in the rat brain. J Comp Neurol 364:523–534
345. Aston-Jones G, Chiang C, Alexinsky T (1991) Discharge of noradrenergic locus coeruleus neurons in behaving rats and monkeys sggests a role in vigilance. Prog Brain Res 88:501–520
346. Loughlin SE, Foote SL, Fallon JJ (1982) Locus coeruleus projections to cortex: topography, morphology and collateralization. Brain Res Bull 9:287–294
347. Loughlin SE, Foote SL, Grzanna R (1986) Efferent projections of nucleus locus coeruleus: morphologic subpopulations have different efferent targets. Neuroscience 18:307–319
348. Loughlin SE, Foote SL, Bloom FE (1986) Efferent projections of nucleus locus coeruleus: topographic organization of cells of origin demonstrated by three-dimensional reconstruction. Neuroscience 18:291–306
349. Haring JH, Davis JN (1985) Differential distribution of locus coeruleus projections to the hippocampal formation: anatomical and biochemical evidence. Brain Res 325:366–369
350. Khoshbouei H, Cecchi M, Dove S, Javors M, Morilak DA (2002) Behavioral reactivity to stress: amplification of stress-induced noradrenergic activation elicits a galanin-mediated anxiolytic effect in central amygdala. Pharmacol Biochem Behav 71:407–417
351. Cecchi M, Khoshbouei H, Morilak DA (2002) Modulatory effects of norepinephrine, acting on 1-receptors in the central nucleus of the amygdala, on behavioral and neuroendocrine responses to acute immobilization stress. Neuropharmacology 43:1139–1147
352. McIntyre CK, Hatfield T, McGaugh JL (2002) Amygdala norepinephrine levels after training predict inhibitory avoidance retention performance in rats. Eur J Neurosci 16:1223–1226
353. Roozendaal B, Koolhaas JM, Bohus B (1993) Posttraining norepinephrine infusion into the central amygdala differentially enhances later retention in Roman high-avoidance and low-avoidance rats. Behav Neurosci 107:575–579
354. Buffalari DM, Grace AA (2007) Noradrenergic modulation of basolateral amygdala neuronal activity: opposing influences of -2 and receptor activation. J Neurosci 27:12358–12366
355. Buffalari DM, Grace AA (2009) Anxiogenic modulation of spontaneous and evoked neuronal activity in the basolateral amygdala. Neuroscience 163:1069–1077
356. Pelosi GG, Tavares RF, Antunes-Rodrigues J, Corrêa FMA (2008) Cardiovascular responses to noradrenaline microinjection in the ventrolateral periaqueductal gray of unanesthetized rats. J Neurosci Res 86:712–719
357. Sullivan GM, Coplan JD, Kent JM, Gorman JM (1999) The noradrenergic system in pathological anxiety: a focus on panic with relevance to generalized anxiety and phobias. Biol Psychiatry 46:1205–1218
358. Southwick SM, Bremner JD, Rasmusson A, Morgan CA III, Arnsten A, Charney DS (1999) Role of norepinephrine in the pathophysiology and treatment of posttraumatic stress disorder. Biol Psychiatry 46:1192–1204
359. Libet B, Gleason CA (1994) The human locus coeruleus and anxiogenesis. Brain Res 634:178–180
360. Blanchard RJ, Taukulis HK, Rodgers RJ, Magee LK, Blanchard DC (1993) Yohimbine potentiates active defensive responses to threatening stimuli in Swiss-Webster mice. Pharmacol Biochem Behav 44:673–681

361. Ivanov A, Aston-Jones G (1995) Extranuclear dendrites of locus coeruleus neurons: activation by glutamate and modulation of activity by alpha-adrenoceptors. J Neurophysiol 74:2427–2436
362. Dierssen M, Gratacòs M, Sahún I, Martin M, Gallego X, Amador-Arjona A, Martinez de Lagran M, Murtra P, Marti E, Pujana MA, Ferrer I, Dalfo E, Martinez-Cue C, Florez J, Torres-Peraza JF, Alberch J, Maldonado R, Fillat C, Estivill X (2006) Transgenic mice overexpressing the full-length neurotrophin receptor TrkC exhibit increased catecholaminergic neuron density in specific brain areas and increased anxiety-like behavior and panic reaction. Neurobiology of Disease 24:403–418
363. Sahún I, Gallego X, Gratacòs M, Murtra P, Trullás R, Maldonado R, Estivill X, Dierssen M (2007) Differential responses to anxiogenic drugs in a mouse model of panic disorder as revealed by Fos immunocytochemistry in specific areas of the fear circuitry. Amino Acids 33:677–688
364. Salchner P, Singewald N (2002) Neuroanatomical substrates involved in the anxiogenic-like effects of acute fluoxetine treatment. Neuropharmacology 43:1238–1248
365. Salchner P, Singewald N (2006) 5-HT receptor subtypes involved in the anxiogenic-like action and associated Fos response of acute fluoxetine treatment in rats. Psychopharmacology 185:282–288
366. Kim M-A, Lee HS, Lee BY, Waterhouse BD (2004) Reciprocal connections between subdivisions of the dorsal raphe and the nuclear core of the locus coeruleus in the rat. Brain Res 1026:56–67
367. Frankhuijzen AL, Wardeh G, Hogenboom F, Mulder AH (1988) Adrenoceptor mediated inhibition of the release of radiolabelled 5-hydroxytryptamine and noradrenaline from slices of the dorsal region of the rat brain. Naunyn-Schmiedeberg's Arch Pharmacol 337:255–260
368. Bortolozi A, Artigas F (2002) Control of 5-hydroxytryptamine release in the dorsal raphe nucleus by the noradrenergic system in rat brain. Role of α-adrenoceptors. Neuropsychopharmacology 28:421–434
369. Hopwood SE, Stamford JA (2001) Noradrenergic modulation of serotonin release in rat dorsal and median raphe nuclei via α-1 and α-2A adrenoceptors. Neuropharmacology 41:433–442
370. Li Y-W, Bayliss DA (1998) Activation of α 2-adrenoceptors causes inhibition of calcium channels but does not modulate inwardly-rectifying K+ channels in caudal raphe neurons. Neuroscience 82:753–765
371. Aston-Jones G (2004) Locus coeruleus, A5 and A7 noradrenergic cell groups. In: Paxinos G (ed) The rat nervous system, 3rd edn. Academic, San Diego, pp 259–294
372. Pickel VM, Joh TH, Reis DJ (1977) A serotonergic innervation of noradrenergic neurons in nucleus locus coeruleus: demonstration by immunocytochemical localization of the transmitter specific enzymes tyrosine and tryptophan hydroxylase. Brain Res 131:197–214
373. Leger L, Descarries L (1978) Serotonin nerve terminals in the locus coeruleus of adult rat: a autoradiographic study. Brain Res 145:1–13
374. Kaehler ST, Singewald N, Phillipu A (1999) Dependence of serotonin release in the locus coeruleus on dorsal raphe neuronal activity. Naunyn-Schmiedeberg's Arch Pharmacol 359:386–393
375. Sinner C, Kaehler ST, Phillipu A, Singewald N (2001) Role of nitric oxide in the stress-induced release of serotonin in the locus coeruleus. Naunyn-Schmiedeberg's Arch Pharmacol 364:105–109
376. Haddjeri N, De Montigny C, Blier P (1997) Modulation of the firing activity of noradrenergic neurones in the rat locus coeruleus by the 5-hydroxytryptamine system. Br J Pharmacol 120:865–875
377. Aston-Jones G, Akaoka H, Charlety P, Chouvet G (1991) Serotonin selectively attenuates glutamate-evoked activation of locus coeruleus neurons in vivo. J Neurosci 11:760–769
378. Aston-Jones G, Shipley MT, Chouvet G, Ennis M, Van Bockastaele EJ, Pieribone VA, Shiekhattar R, Akaoka H, Drolet G, Astier B, Charlety P, Valentino R, Williams JT (1991) Afferent regulation of locus coeruleus neurons: anatomy, physiology and pharmacology. Prog Brain Res 88:47–75

Chapter 4
The Deakin–Graeff Hypothesis

The first reports on the role of the serotonergic system in the control of defensive behavior in animal models date from the 1970s. Robichaud and Sledge [1] demonstrated that DL-para-chloropheynalainine (DL-pCPA), a serotonin synthesis inhibitor, releases punished behavior. Shortly after that, Graeff and Schoenfeld [2, 3] demonstrated that metisergide, lysergic acid, and bromolysergic acid (non-selective antagonists at 5-HT receptors) produce similar effects, while the non-selective agonist α-methyltryptamine increased the effect of punishment. In the same direction, the destruction of serotonergic fibers in the ventromedial tegmentum of rats has been shown to prevent the acquisition of conflict-induced response suppression [4]. Shortly after that, Wise and colleagues reported that the benzodiazepine oxazepam decreases serotonin turnover in the midbrain in the same dose that releases punished behavior [5]. This effect is probably due to benzodiazepine modulation of $\alpha_3\beta_n\gamma_2$ $GABA_A$ receptors, which are highly expressed in serotonergic neurons of the raphe [6] and seems to be responsible for the anxiolytic effect of benzodiazepines [7, 8]. The target of the serotonergic projections involved in these experiments seem to be the amygdala, since the microinjection of serotonin antagonists in the basolateral nucleus (BLA) produces an anti-conflict effect (i.e., release behavior that was suppressed by punishment superimposed on positive reinforcement) [9].

Serotonin seems to have an opposite role in the periaqueductal gray area, given that manipulations which decrease serotonergic activity (such as the administration of DL-pCPA or non-selective 5-HT antagonists) increase bar press responses which turn off the electrical stimulation of this region, and treatments which increase serotonergic activity reduce this responding [10–13]. The microinjection of the $5\text{-}HT_{1A}R$ full agonist 8-OH-DPAT in the dorsal periaqueductal gray (dPAG) decreases escape responses elicited by stimulation of the dPAG with homocysteic acid [14]. The intra-dPAG administration of serotonergic antagonist by itself does not present effects on excitatory amino acid (EAA)-elicited escape, but prevents the effect of agonists [15, 16], suggesting that serotonergic projections to the dPAG exert a phasic inhibition in defensive behavior controlled by the PAG. Deakin and Graeff [17] suggested that the modulatory influence of serotonin on

C. Maximino, *Serotonin and Anxiety*, SpringerBriefs in Neuroscience,
DOI: 10.1007/978-1-4614-4048-2_4, © The Author(s) 2012

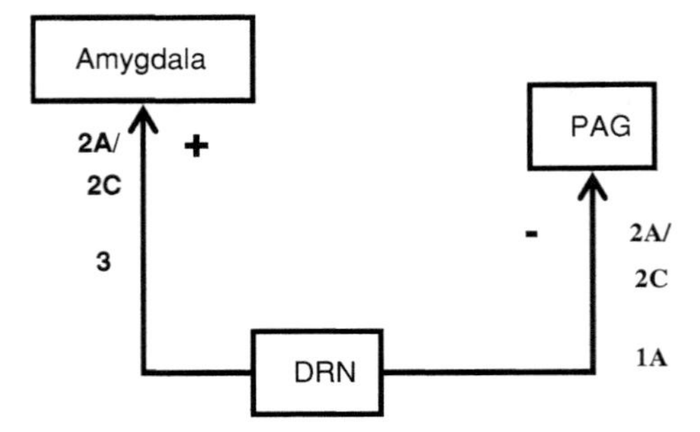

Fig. 4.1 The dual role of serotonin in defense responses mediated by amygdala or periaqueductal gray area (*PAG*). In the amygdala, serotonin released from dorsal raphe nucleus (*DRN*) projections increase anxiety- or fear-like behavior via 2A/2C or 3 receptors. In the periaqueductal gray, activation of 2A/2C or 1A receptors decreases anxiety- or fear-like behavior

PAG functioning manifests itself only in situations which engage the serotonergic system.

The observation of differential effects of serotonin on amygdala- and PAG-mediated defensive responses led Deakin and Graeff [17] to postulate that this neurotransmitter exerts opposing actions on anxiety-like behavior—putatively mediated by amygdaloid nuclei (via 5-HT_2 and 5-HT_3 receptors)—and fear-like behavior, putatively mediated by the periaqueductal gray (via 5-HT_{1A} and 5-HT_2 receptors), with serotonin increasing the former and inhibiting the later. This is the "dual role" Deakin–Graeff hypothesis of serotonin function in anxiety (Fig. 4.1).

A little after the enunciation of the Deakin–Graeff hypothesis, new ethologically inspired tests of anxiety-like behavior were introduced, including the elevated plus-maze (EPM), which quickly became the "gold standard" in anxiety research. However, serotonergic agents produced contradictory or inconclusive results in this test, leading to a closer observation of the behavioral repertoire of animals in these tests [18]. For example, Sheila Handley argued that rat behavior in the EPM reflects two defensive strategies that could be influenced in opposite directions by 5-HT, open-arm avoidance when the animal is in one of the closed arms (inhibitory avoidance), and escape from an open-arm in the direction of the enclosed arm (one-way escape) [19].

4.1 Destruction or Blockade of DRN Neurons is Anxiolytic and Panicogenic

The inactivation of the dorsal raphe nucleus, either by neurotoxic lesions or by the injection of GABA receptor agonists, tends to produce anxiolytic effects. The injection of muscimol in the DRD or of the serotonergic neurotoxin 5,7-dihydroxytryptamine (5,7-DHT) in the DRV facilitates one-way escape and impairs inhibitory avoidance in the elevated T-maze (ETM) [20] and is anxiolytic in the social interaction test [21]. Consistent with these anxiolytic effects, 8-OH-DPAT microinjections in the same region release punished responding in the Geller-Seifter conflict model [22] and is anxiolytic in the ETM and in trial 2 of the EPM [23], while injections of WAY 100635 have opposite effects [24].

4.2 The Defensive Context for Increased Serotonin Release

Serotonin release in the limbic forebrain is dependent on aspects of the environment which compose a "defensive context". These environmental aspects are comparable to those which have been discussed in Sects. 1.1 and 1.2. Thus, conflict is an important aspect for serotonin release in hippocampal regions. While exposure to a normal EPM increases serotonin release in the ventral hippocampus [25, 26], an "inactivated" maze (i.e., a maze with no open arms) or exposure to white noise in the home cage does not [26]. Likewise, rats subjected to the Vogel conflict test show increased 5-HT release in the dorsal hippocampus [27]. Exposure to an EPM [28] and performance of avoidance (but not escape) in an ETM [29] increase c-Fos-like immunoreactivity in the dorsal raphe nucleus, as does exposure to an open field under high illumination conditions [30].

Stressor controllability also seems to be an important aspect for the "defensive context" of serotonin release. (Uncontrollable) restraint stress increases serotonin release in the central amygdala [31], dorsomedial hypothalamus [32], and frontal cortex [33]. This latter effect is potentiated by local application of the citalopram, and the effect of this SSRI is inhibited by pretreatment with the 5-HT_{2C}R agonist RO 60-0175 [33]. Inescapable, but not escapable, shock increases 5-HT release in the BLA and ventral hippocampus, and rats exposed to previously inescapable shocks exhibit exaggerated 5-HT responses to two brief footshocks [34, 35]. c-Fos-like immunoreactivity is also observed in the DRN after inescapable, but not escapable, shock [36]. Exposure of mice to a predator in an inescapable situation increases serotonin release in the hippocampus, medial prefrontal cortex, and septum, but not striatum, of mice [37]. So far, it is not known whether the availability (and performance) of escape routes during predator exposure also affects serotonin release in these regions.

These effects are mediated by peptidergic mechanisms in the raphe. Microinjection of CRF in the DRN of rats induce freezing accompanied by an increase in serotonin release in the CeA; in contrast, the cessation of freezing is accompanied by a prolonged increase in 5-HT release in the mPFC [38]. This last effect is blocked by the administration of the CRF_2R antagonist antisauvagine-30, but not the CRF_1R antagonist antalarmin [39]. Likewise, exposure of guinea pigs to an EPM increases serotonin release in the prefrontal cortex, an effect which is potentiated by cholecystokinin receptor B (CCK-B) agonists [40–42]; these drugs also produce a marked anxiogenic profile. CCK-B agonists do not affect anxiety-like behavior or 5-HT levels when animals are observed in the home cage [40], and

Table 4.1 Anxiogenic and anxiolytic neuropeptides, as well as anxiogenic and anxiolytic drugs, have predictable effects on serotonin release. Adapted from Ref. [44]

Substance	Behavior	5-HT
Corticotropin-releasing factor (CRF)	Anxiogenic	↑ Hippocampus, amygdala, prefrontal cortex ↓ Lateral septum
Cholecystokinin (CCK)	Anxiogenic	↑ Prefrontal cortex
Substance P	Anxiogenic	↑ Hippocampus
Galanin	Anxiolytic	↓ Hippocampus
Hypocretin/orexin	Anxiogenic	↑ Dorsal raphe
Urocortin 1, 2	Anxiogenic	↑ Hippocampus, amygdala
Somatostatin (SST)	Anxiolytic	↑ Hippocampus
Adenosine A_1R agonists	Anxiolytic	↓ Hippocampus
Cannabinoid CB_1R agonists	Anxiolytic	↓ Hippocampus, prefrontal cortex
Nitric oxide donors	Biphasic	Biphasic dorsal raphe, hypothalamus, hippocampus, prefrontal cortex
Benzodiazepines	Anxiolytic	↓Hippocampus
Triazolo-benzodiazepines	Anxiolytic	↑ Hippocampus

their effect is blocked by systemic administration of an anxiolytic dose of 8-OH-DPAT [41] or diazepam [43], suggesting that CCK-B receptors mediate changes in 5-HT release under aversive conditions, but not in resting states.

Overall, mildly-to-highly anxiogenic situations tend to increase activity in the dorsal raphe nucleus with concomitant serotonin release in forebrain targets of the DRN. Moreover, anxiogenic peptides lead to an increase in serotonin levels in the hippocampus, amygdala, and prefrontal cortex, while anxiolytic peptides have the opposite effect (with the possible exception of somatostatin) [44]. Likewise, other neurotransmitter systems which are thought to mediate anxiety-like behavior (adenosine [45]; nitric oxide [46]; endocannabinoids [47, 48]) directly modulate serotonin release in anxiogenic situations (Table 4.1).

References

1. Robichaud RC, Sledge KL (1969) The effects of p-chlorophenylalanine on experimentally induced conflict in the rat. Life Sci 8:965–969
2. Graeff FG, Schoenfeld RI (1970) Tryptaminergic mechanisms in punished and nonpunished behavior. J Pharmacol Exp Ther 173:277–283
3. Schoenfeld RI (1976) Lysergic acid diethylamide- and mescaline-induced attenuation of the effect of punishment in the rat. Science 192:801–803
4. Tye NC, Everitt BJ, Iversen SD (1977) 5-Hydroxytryptamine and punishment. Nature 268:741–742
5. Wise CD, Berger BD, Stein L (1972) Benzodiazepines: anxiety-reducing activity by reduction of serotonin turnover in the brain. Science 177:180–183
6. Gao B, Fritschy JM, Benke D, Mohler H (1993) Neuron-specific expression of GABAA-receptor subtypes: differential association of the 1- and 3-subunits with serotonergic and gabaergic neurons. Neuroscience 54:881–892

7. Dias R, Sheppard WFA, Fradley RL, Garrett EM, Stanley JL, Tye SJ, Goodacre S, Lincoln RJ, Cook SM, Conley R, Hallett D, Humphries AC, Thompson SA, Wafford KA, Street LJ, Castro JL, Whiting PJ, Rosahl TW, Atack JR, McKernan RM, Dawson GR, Reynolds DS (2005) Evidence for a significant role of 3-containing GABAA receptors in mediating the anxiolytic effects of benzodiazepines. J Neurosci 25:10682–10688
8. Atack JR, Hutson PH, Collinson N, Marshall G, Bentley G, Moyes C, Cook SM, Collins I, Wafford K, McKernan RM, Dawson GR (2005) Anxiogenic properties of an inverse agonist selective for 3 subunit-containing GABAA receptors. Br J Pharmacol 144:357–366
9. Hodges H, Green S, Glenn B (1987) Evidence that the amygdala is involved in benzodiazepine and serotonergic effects on punished responding but not in discrimination. Psychopharmacology 92:491–504
10. Kiser RS Jr, Lebovitz RM (1975) Monoaminergic mechanisms in aversive brain stimulation. Physiol Behav 15:47–53
11. Kiser RS, Lebovitz RM, German DC (1978) Anatomic and pharmacologic differences between two types of aversive midbrain stimulation. Brain Res 155:331–342
12. Schenberg LC, Graeff FG (1978) Role of the periaqueductal gray substance in the antianxiety action of benzodiazepines. Pharmacol Biochem Behav 9:287–295
13. Kiser RS, German DC, Lebovitz RM (1978) Serotonergic reduction of dorsal central gray area stimulation-produced aversion. Pharmacol Biochem Behav 9:27–31
14. Beckett S, Marsden CA (1997) The effect of central and systemic injection of the 5-HT1A receptor agonist 8-OHDPAT and the 5-HT1A receptor antagonist WAY 100635 on periaqueductal grey-induced defence behaviour. J Psychopharmacol 11:35–40
15. Nogueira RL, Graeff FG (1991) Mediation of the antiaversive effect of isamoltane injected into the dorsal periaqueductal grey. Behav Pharmacol 2:73–77
16. Schütz MTB, Aguiar JC, Graeff FG (1985) Anti-aversive role of serotonin on the dorsal periaqueductal grey matter. Psychopharmacology 85:340–345
17. Deakin JFW, Graeff FG (1991) 5-HT and mechanisms of defence. J Psychopharmacol 5:305–315
18. Rodgers RJ (1997) Animal models of 'anxiety': Where next? Behav Pharmacol 8:477–496
19. Handley SL, McBlane JW, Critchley MAE, Njung'e K (1993) Multiple serotonin mechanisms in animal models of anxiety: environmental, emotional and cognitive factors. Behav Brain Res 58:203–210
20. Sena LM, Bueno C, Pobbe RLH, Andrade TGCS, Zangrossi H Jr, Viana MB (2002) The dorsal raphe nucleus exerts opposed control on generalized anxiety and panic-related defensive responses in rats. Behav Brain Res 142:125–133
21. File SE, Hyde JRG, MacLeod NK (1979) 5,7-Dihydroxytryptamine lesions of dorsal and median raphé nuclei and performance in the social interaction test of anxiety and in a home-cage aggression test. J Affect Disord 1:115–122
22. Cervo L, Mocaër E, Bertaglia A, Samanin R (2000) Roles of 5-HT1A receptors in the dorsal raphe and dorsal hippocampus in anxiety assessed by the behavioral effects of 8-OH-DPAT and S 15535 in a modified Geller-Seifter conflict model. Neuropharmacology 39:1037–1043
23. File SE, Gonzalez LE (1996) Anxiolytic effects in the plus-maze of 5-HT1A-receptor ligands in dorsal raphé and ventral hippocampus. Pharmacol Biochem Behav 54:123–128
24. Pobbe RLH, Zangrossi H Jr (2005) 5-HT1A and 5-HT2A receptors in the rat dorsal periaqueductal gray mediate the antipanic-like effect induced by the stimulation of serotonergic neurons in the dorsal raphe nucleus. Psychopharmacology 183:314–321
25. Wright IK, Upton N, Marsden CA (1992) Effect of established and putative anxiolytics on extracellular 5-HT and 5-HIAA in the ventral hippocampus of rats during behaviour on the elevated X-maze. Psychopharmacology 109:338–346
26. Rex A, Voigt JP, Fink H (2005) Anxiety but not arousal increases 5-hydroxytryptamine release in the rat ventral hippocampus in vivo. Eur J Neurosci 22:1185–1189
27. Matsuo M, Kataoka Y, Mataki S, Kato Y, Oi K (1996) Conflict situation increases serotonin release in rat dorsal hippocampus: in vivo study with microdialysis and Vogel test. Neurosci Lett 215:197–200

28. Silveira MC, Sandner G, Graeff FG (1993) Induction of Fos immunoreactivity in the brain by exposure to the elevated plus-maze. Behav Brain Res 45:115–118
29. Silveira MCL, Zangrossi H Jr, Viana MB, Silveira R, Graeff FG (2001) Differential expression of Fos protein in the rat brain induced by performance of avoidance or escape in the elevated T-maze. Behav Brain Res 126:13–21
30. Hale MW, Hay-Schmidt A, Mikkelsen J, Poulsen B, Bouwknecht JA, Evans AK, Stamper CE, Shekhar A, Lowry CA (2008) Exposure to an open-field arena increases c-Fos expression in a subpopulation of neurons in the dorsal raphe nucleus, including neurons projecting to the basolateral amygdaloid complex. Neuroscience 157:733–748
31. Mo B, Feng N, Renner KJ, Forster GL (2008) Restraint stress increases serotonin release in the central nucleus of the amygdala via activation of corticotropin-releasing factor receptors. Brain Res Bull 76:493–498
32. Lowry CA, Plant A, Shanks N, Ingram CD, Lightman SL (2003) Anatomical and functional evidence for a stress-responsive, monoamine-accumulating area in the dorsomedial hypothalamus of adult rat brain. Horm Behav 43:254–262
33. Mongeau R, Martin CBP, Chevarin C, Maldonado R, Hamon M, Robledo P, Lanfumey L (2010) 5-HT2C receptor activation prevents stress-induced enhancement of brain 5-HT turnover and extracellular levels in the mouse brain: modulation by chronic paroxetine treatment. J Neurochem 115:438–449
34. Amat J, Matus-Amat P, Watkins LR, Maier SF (1998) Escapable and inescapable stress differentially alter extracellular levels of 5-HT in the basolateral amygdala of the rat. Brain Res 812:113–120
35. Amat J, Matus-Amat P, Watkins LR, Maier SF (1998) Escapable and inescapable stress differentially and selectively and alter extracellular levels of 5-HT in the ventral hippocampus and dorsal periaqueductal gray of the rat. Brain Res 797:12–22
36. Grahn RE, Will MJ, Hammack SE, Maswood S, McQueen MB, Watkins LR, Maier SF (1999) Activation of serotonin-immunoreactive cells in the dorsal raphe nucleus in rats exposed to an uncontrollable stressor. Brain Res 826:35–43
37. Beekman M, Flachskamm C, Linthorst ACE (2005) Effects of exposure to a predator on behaviour and serotonergic neurotransmission in different brain regions of C57bl/6N mice. Eur J Neurosci 21:2825–2836
38. Forster GL, Feng N, Watt MJ, Korzan WJ, Mouw NJ, Summers CH, Renner KJ (2006) Corticotropin-releasing factor in the dorsal raphe elicits temporally distinct serotonergic responses in the limbic system in relation to fear behavior. Neuroscience 141:1047–1055
39. Forster GL, Pringle RB, Mouw NJ, Vuong SM, Watt MJ, Burke AR, Lowry CA, Summers CH, Renner KJ (2008) Corticotropin-releasing factor in the dorsal raphe nucleus increases medial prefrontal cortical serotonin via type 2 receptors and median raphe nucleus activity. Eur J Neurosci 28:299–310
40. Rex A, Fink H (1998) Effects of cholecystokinin-receptor agonists on cortical 5-HT release in guinea pigs on the X-maze. Peptides 19:519–526
41. Rex A, Marsden CA, Fink H (1997) Cortical 5-HT-CCK interactions and anxiety-related behaviour of guinea-pigs: a microdialysis study. Neurosci Lett 228:79–82
42. Rex A, Fink H, Marsden CA (1994) Effects of BOC-CCK-4 and L 365.260 on cortical 5-HT release in guinea-pigs on exposure to the elevated plus maze. Neuropharmacology 33:559–565
43. Rex A, Marsden CA, Fink H (1993) Effect of diazepam on cortical 5-HT release and behaviour in the guinea-pig on exposure to the elevated plus-maze. Psychopharmacology 110:490–496
44. Steckler T (2008) Peptide receptor ligands to treat anxiety disorders. In: Blanchard RJ, Blanchard DC, Griebel G, Nutt D (eds) Handbook of anxiety and fear. Elsevier B. V, Amsterdam, pp 157–221
45. Florio C, Prezioso A, Papaioannou A, Vertua R (1998) Adenosine A1 receptors modulate anxiety in CD1 mice. Psychopharmacology 136:311–319

46. Zhang J, Huang X-Y, Ye M-L, Luo C-X, Wu H-Y, Hu Y, Zhou Q-G, Wu D-L, Zhu L-J, Zhu D-Y (2010) Neuronal nitric oxide synthase alteration accounts for the role of 5-HT1A receptor in modulating anxiety-related behaviors. J Neurosci 30:2433–2441
47. Haj-Dahmane S, Shen R-Y (2009) Endocannabinoids suppress excitatory synaptic transmission to dorsal raphe serotonin neurons through the activation of presynaptic CB1 receptors. J Pharmacol Exp Ther 331:186–196
48. Haj-Dahmane S, Shen R-Y (2011) Modulation of the serotonin system by endocannabinoid signaling. Neuropharmacology (in press)

Chapter 5
Topographic Organization of DRN

While anxiogenic stimuli activate neurons in the raphe and lead to serotonin release in limbic forebrain targets, panicogenic stimuli do not necessarily do so. For example, escape performance in an elevated T-maze does not increase c-Fos-like immunoreactivity in the raphe [1]. Nonetheless, the observation of a panicolytic role of serotonin in the PAG and an anxiogenic role in the amygdala and hippocampus suggests that the raphe is not a homogeneous structure. In fact, the dorsal raphe can be divided at least into six subregions based on cytoarchitecture and distribution of serotonergic neurons [2–4]. These comprise the rostral (DRr), dorsal (DRD), ventral (DRV), lateral wing (lwDR), caudal (DRC), and interfascicular (DRI) portions (Fig. 5.1). Among those, the rostral, ventral, and interfascicular subregions play little role in the control of defense responses, and discussion of their functions can be found elsewhere [2, 3]. Here, we will discuss evidence for a role of the DRD, lwDR and DRC in anxiety and fear.

5.1 The Dorsal Portion of the DRN is Part of a Mesocorticolimbic System Involved in Anxiety-Like Responses

The DRD is a tryptophan hydroxylase-rich group of small-sized, fusiform, or bipolar cells that extends from the caudal border of the trochlear nuclei to the rostral border of the dorsal tegmental nuclei [3, 5, 6]. Projections to this subregion come from the mPFC, CeA, BNST, LH, and DMH [3, 5]. This region presents moderate levels of tryptophan hydroxylase 2, serotonin transporter, organic cation transporter 3, plasma membrane monoamine transporter, 5-HT_{1A} and 5-HT_{1B} receptors [7–10]. Among all subregions, serotonergic neurons from the DRD show the lowest resting membrane potential, the highest activation gap (i.e., the difference between the resting membrane potential and the action potential threshold), and the smaller response to 5-CT [6]. In relation to neurons from the

C. Maximino, *Serotonin and Anxiety*, SpringerBriefs in Neuroscience,
DOI: 10.1007/978-1-4614-4048-2_5, © The Author(s) 2012

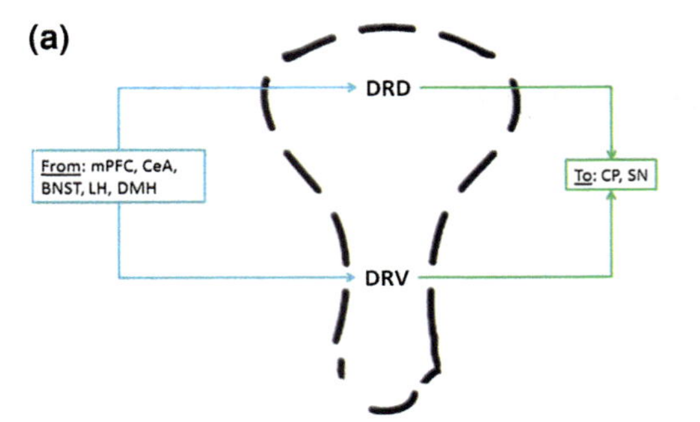
(a)
DRD
From: mPFC, CeA, BNST, LH, DMH
To: CP, SN
DRV
Bregma -7.30 mm

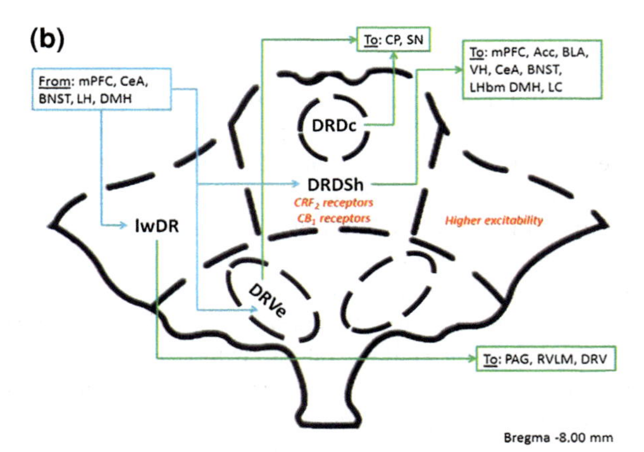
(b)
To: CP, SN
To: mPFC, Acc, BLA, VH, CeA, BNST, LHbm DMH, LC
From: mPFC, CeA, BNST, LH, DMH
DRDc
DRDSh
CRF2 receptors
CB1 receptors
Higher excitability
lwDR
DRVe
To: PAG, RVLM, DRV
Bregma -8.00 mm

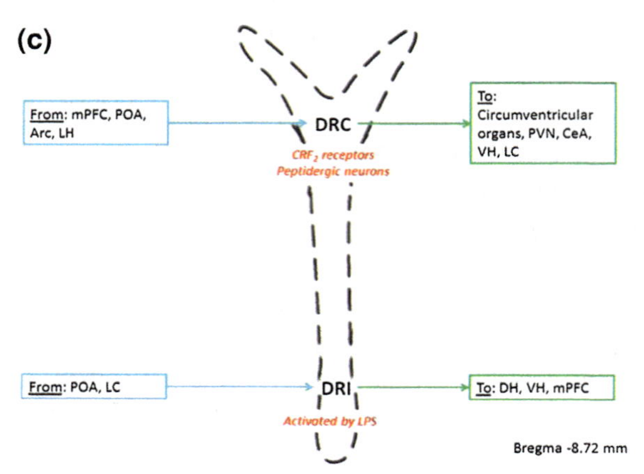
(c)
From: mPFC, POA, Arc, LH
DRC
To: Circumventricular organs, PVN, CeA, VH, LC
CRF2 receptors
Peptidergic neurons
From: POA, LC
DRI
To: DH, VH, mPFC
Activated by LPS
Bregma -8.72 mm

◀ **Fig. 5.1** Topographical organization of projections from and to the DRN, as well as physiological and pharmacological properties of each area. **a** Midline portion of the raphe (at bregma −7.3 mm in rats), discriminating a dorsal portion (DRD) and a ventral portion (DRV). **b** Midline-to-caudal portion (at bregma −8 mm in rats), showing both divisions of the DRD (shell [DRDSh] and core [DRDc]), lateral wings (lwDR), and a subdivision of the DRV, the "ellipses" (DRVe). **c** Caudal-most raphe nucleus (at bregma −8.72 mm in rats) demonstrating the caudal portion (DRC) and the interfascicular portion (DRI) of the dorsal raphe nucleus. Abbreviations: Acc, nucleus accumbens; Arc, arcuate nucleus; BNST, bed nucleus of the stria terminalis; CeA, central amygdala; CP, caudate-putamen; DH, dorsal hippocampus; DMH, dorsomedial hypothalamus; DRC, caudal portion of the dorsal raphe nucleus; DRD, dorsal portion of the dorsal raphe nucleus; DRDc, core of the dorsal portion of the dorsal raphe nucleus; DRDSh, shell of the dorsal portion of the dorsal raphe nucleus; DRI, interfascicular portion of the dorsal raphe nucleus; DRV, ventral portion of the dorsal raphe nucleus; DRVe, ellipses of the ventral portion of the dorsal raphe nucleus; LC, locus coeruleus; LH, lateral hypothalamus; LHbm, medial portion of the lateral habenula; LPS, lipopolysaccharide; lwDR, lateral wings of the dorsal raphe nucleus; mPFC, medial prefrontal cortex; PAG, periaqueductal gray area; POA, preoptic area;PVN, paraventricular nucleus; RVLM, rostroventrolateral medulla; SN, substantia nigra; VH, ventral hippocampus

ventral DRN (DRV), the dendritic arbor of DRD cells is widely dispersed, extending into all directions [6].

An important neurochemical property of the DRD is the expression of CRF in serotonergic neurons [11]; these neurons are surrounded by a dense plexus of fibers which express the neurokinin 1 receptor (NK_1R) [12]. Intracerebroventricular CRF injections inhibit ~45 % of these neurons and excite ~30 % [13]. These CRF-containing neurons project to CRFergic neurons in the central amygdala [11]. Interestingly, CRF inhibits activity of CeA neurons in vitro [12] and serotonin inhibits the activity of a subset of neurons located in the lateral CeA, "Type I" interneurons [14, 15]. Thus, CRF from the DRD may act as a co-transmitter with serotonin, fine-tuning the activity of CeA neurons [16].

Moderate levels of the CRF_2 receptor are also observed in serotonergic neurons of the DRD [17]. Activation of these receptors by urocortin II at low doses produces a short-lived inhibition of non-serotonergic neurons, while higher doses excited serotonergic neurons [18]. Additionally, mice lacking the CRF_2R fail to display elevations in TPH2 and CRF_1R levels in the DRD and in anxiety-like behavior after chronic unpredictable stress; however, they also fail to increase antiapoptotic gene expression in the DRD and lwDR after chronic stress, presenting significant cell loss in these regions after stress [19].

In the DRD, CB_1 endocannabinoid receptor expression is considerably high, with about 35 % of all serotonergic neurons presenting CB_1R mRNA [20]. Intraraphe injections of cannabinoid agonists produce an antinociceptive effect [21]. In DRN slices, CB_1R antagonists reduce firing of serotonergic neurons through a $GABA_AR$-dependent mechanism [22], while CB_1R agonists reduce 5-HTergic neuron activity in a manner consistent with decreased glutamate release to those cells [23]. Systemic administration of CB_1R antagonists also increase serotonin release in the medial prefrontal cortex, nucleus accumbens, and hippocampus [24–28] and is anxiogenic in the elevated plus-maze, open field, and elevated T-maze [28], all regions targeted by DRD projections [5]. In the elevated plus-maze and hole-board tests, the non-selective cannabinoid receptor agonist CP55940 is anxiolytic at low doses and anxiogenic at high doses; the latter effect is attenuated by systemic administration of the 5-HT_{1A}R antagonist WAY 100635 [29].

The DRD can be further subdivided in a "core" (DRDc) and a "shell" (DRDSh), with the core presenting a compact cluster of serotonergic neurons and the shell presenting scattered serotonergic neurons [2, 3, 5]. Neurons from the DRDc project to the interstitial nucleus of Cajal, nucleus of fields of Forel, caudate-putamen, and parieto-temporal cortex, while cells from the DRDSh project to the habenula, dorsal hypothalamus, globus pallidus, amygdala, BNST, nucleus accumbens, entorrhinal cortex, hippocampus, piriform cortex, medial prefrontal cortex, and olfactory tubercles [11, 30–46]. A high density of GABA transporter 1 (GAT1)-positive fibers terminate in the shell region, suggesting that projections to the DRDSh are GABAergic, or at least accumulate GABA [6]. However, since the dendritic arbor of DRD neurons is widely dispersed, it is possible that cells from the core also receive GABAergic input [3, 5].

Serotonergic cells of the DRDSh usually co-express vesicular glutamate transporter 3 (vGlut3) [6]; the non-serotonergic cells of the DRDSh can also express vGlut3, and project to the ventral tegmental area, substantia nigra pars compacta, anterior hypothalamic area, and preoptic area [47], while neurons projecting to the BLA are mostly serotonergic [48]. The non-serotonergic neurons usually express 5-HT_{1A} receptors but respond with very little hyperpolarization to the non-selective agonist 5-CT [6, 49]; they are also the only neurons of the DRN to show shorter afterhyperpolarization than non-serotonergic neurons [6], and do not express nitric oxide synthase [50]. Some of these neurons co-express glutamic acid decarboxylase (GAD) and α_{1b} and/or 5-HT_{1A} receptors, but not CRF receptors [17]. Serotonergic neurons in the DRDc and DRDSh, on the other hand, show a gradual, slow after hyperpolarization and short membrane time constant [49], and usually co-express nitric oxide synthase [50]. The role of nitric oxide in DRD neurons has not been determined. Nonetheless, injection of the nitric oxide synthase inhibitor L-NAME in the DRN blocks the acquisition of learned helplessness in inescapable shock-administered rats [51], and exposure to an elevated plus-maze or to a live cat activates NADPH-diaforase-positive cells in the DRN [52, 53]. Nitric oxide does not seem to be co-released with 5-HT in projection areas from the DRDc, since nNOS is not localized in axon terminals in the striatum, and should have a local regulatory role in serotonin levels [54].

As mentioned above, exposure of rats to an open field activates neurons in the DRD which project to the BLA [48]. These cells are serotonergic neurons located in the DRDSh [48]. The administration of anxiogenic drugs activate neurons in the DRD which project to the BLA [38] and parvalbumin- and 5-HT_{2A}-R-expressing GABAergic neurons in the BLA [55]. Fear-potentiated startle, on the other hand, activates neurons in the DRD and in the CeA [56]; these latter neurons are unlikely to be "Type I" neurons, which are inhibited by serotonin [14, 15]. The connectivity pattern of the DRD, together with the c-Fos data, strongly suggests that this subregion regulates anxiety-like behavioral and physiological responses [2, 3, 11, 31, 57].

5.2 The Caudal Portion of the DRN is Highly Responsive to Stress-Related Peptides

The caudal portion of the DRN (DRC) is located in the rostral pons, bordering the dorsal tegmental nucleus [5, 11], and contains small round serotonergic neurons in its rostral part and medium-sized cells [3]. The GABAergic and serotonergic neurons of the DRC present a high density of CRF_2 receptors [17]. It has been reported that while nNOS is weakly expressed in the DRC [50], the co-expression of nNOS with 5-HT and galanin in this subregion is higher than in the DRD [54]. DRC neurons also present CCKergic and enkephalinergic neurons [58, 59] and fibers immunoreactive to calcitonin gene-related peptide (CGRP), CCK, dynorphin B, neuropeptide Y, somatostain, and substance P [60, 61].

The vesicular glutamate transporter vGlut3 is also expressed in serotonergic neurons of the DRC which project to the forebrain [47, 62]. vGlut3 expression seems to be independent of Lmx1b [63], a transcription factor that specifies the serotonergic phenotype in parallel with Pet-1. Interesting findings regarding the serotonergic system in vGlut3 knockout mice have been reported by Amilhon and colleagues [64]. These animals show impairments in 5-HT_{1A} autoreceptor function in the DRN, with decreased inhibition of firing by ipsapirone, decreased coupling to G-proteins in response to 5-CT, and decreased loss of body temperature after systemic administration of 8-OH-DPAT [64]. Accompanying these presynaptic effects, the glutamate-induced, reserpine-sensitive vesicular uptake of serotonin in cortex was abolished in *vglut3*$^{-/-}$ mice. In the ventral hippocampus, paroxetine-stimulated (but not basal) 5-HT release is decreased in knockout animals, and serotonin turnover is increased. An important effect of the genetic ablation of vGlut3 is a selective effect on anxiety-like behavior, as these animals show increased open-arm avoidance in an elevated plus-maze, increased latency to feed in the novelty-suppressed feeding test, and faster marble burying. In *vglut3*$^{-/-}$ mice (but not in wild-type littermates), the increased latency in the novelty-suppressed feeding test is abolished after treatment with the tryptophan hydroxylase inhibitor pCPA. In pups on postnatal age P8, maternal separation leads to longer ultrasonic vocalization bouts in knockouts, suggesting that the increased anxiety in these animals is established at an early developmental stage [64].

The DRC receives massive glutamatergic inputs from the medial prefrontal cortex and from hypothalamic regions (preoptic area, arcuate nucleus, and perifornical and lateral hypothalamus) [65]. Glutamatergic projections from the cortex express mainly vGlut1 and terminate on distal dendrites; subcortical glutamatergic projections, on the other hand, express mainly vGlut2 and terminate on proximal dendrites and cell bodies [61]. Projections from the DRC concentrate mainly on circumventricular organs and ependymal cells lining the ventricles [66–68], and on the ventral hippocampus, amygdala, and locus coeruleus [30]. A subset of serotonergic neurons from the DRC discharge action potentials is phase-locked to the hippocampal theta rhythm [69]. As argued, the slow clock-like activity that is attributed to "putative serotonergic" cells [70] has an important role in maintaining state-dependent stable serotonin levels in the raphe and its projections; rhythmically synchronized firing, on the other hand, could provide more precise timing in serotonin release to targets which are also oscillating, including the hippocampus, cortex, and amygdala [69].

A portion of the raphe located between the DRD and DRC (midline-to-caudal DR) forms a specialized region of the DRC. Intracerebroventricular injections of urocortin 2 increase c-Fos expression in serotonergic neurons of this region [71, 72]; 35 % of the activated neurons project to the periventricular system [73]. This subdivision also projects to the parvocellular subdivision of the paraventricular nucleus, lateral parabrachial nucleus, and central amygdala [74, 75], suggesting an important role for the DRC in modulating the autonomic nervous system and anxiety-related circuits [76].

Neurons in DRC are sensitive to anxiogenic drugs and anxiety-related neuropeptides [2, 3, 38, 77]. Audiogenic stress increases tryptophan hydroxylase activity in the DRC [78], a sine qua non condition for increased firing rate in these neurons [79]. The DRC has been implicated in the behavioral effects of inescapable stress (deficits in escape acquisition and increases in fear conditioning). c-Fos-like immunoreactivity and 5-HT efflux are increased in the DRC after inescapable, but not escapable, shocks [80, 81]. Likewise, overexpression of 5-HT_{1B} receptors in the DRC, but not in the rostral DRN, reduces immobility in the forced swim test [82, 83].

The observation that CRF increases the in vitro activity of tryptophan hydroxylase in the DRC suggests a role for this peptide in the control of serotonin activity in this subregion [78]. Indeed, blockade of CRF receptors in the DRC, but not DRr, blocks the sustained anxiogenic effect of inescapable stress [84]. Conversely, microinjection of urocortin 2 into the DRC mimics the behavioral effects of inescapable stress [85], an effect which is mediated by GABAergic interneurons in the DN [86]. It has been proposed that, while CRF_1 receptors mediate the direct or indirect inhibition of DRC 5-HT neurons, thus promoting active coping, CRF_2 receptors mediate the disinhibition of these cells, promoting passive coping [87].

Another system which has been implicated in the control of DRC activity during the establishment of learned helplessness is the nitrergic system. The microinjection of an NMDA antagonist on the DRC during inescapable shock prevents its effects on fear conditioning and escape acquisition. Injection 24 h later, before testing, reduces learned helplessness, but does not block it. Conversely, microinjection of a nitric oxide synthase inhibitor blocks learned helplessness when injected before testing, and reduces it when injected during inescapable shock [88]. The source of this glutamatergic input to the DRC is unknown. One area which has been implicated is the lateral habenula, as excitatory projections from the LHb synapse on GABAergic and serotonergic neurons of the dorsal raphe nucleus [89, 90]. Habenular lesions blocks learned helplessness and the differential increase in extracellular levels of 5-HT on the DRC [91], suggesting that this structure is necessary for the behavioral and physiological effects of inescapable shocks. A similar behavioral effect is observed in pharmacogenetic lesions of the habenula of larval zebrafish [92].

While an excitatory projection from the LHb blocks the consequences of inescapable stress, a direct or indirect projection from the medial prefrontal cortex also modulates the DRC during learned helplessness. Microinjections of muscimol into the prelimbic and infralimbic cortices (but not in cingulate or ventral orbital cortices) of rats increases escapable shock-induced c-Fos-like immunoreactivity in serotonin neurons and extracellular 5-HT concentration in the DRC up to the levels which are produced by inescapable shock; interestingly, these animals also behave as if they have been previously exposed to inescapable shocks [93]. Importantly, muscimol microinjections had no effect on behavior or physiology of animals exposed to inescapable shocks. It is probable that stressor controllability is assessed by the prelimbic and infralimbic cortices, which inhibits (through activation of GABAergic interneurons) the DRC when events are controllable.

5.3 The Lateral Wings of the DRN are Involved in Panic-Like Responses

The lateral wings of the DRN (lwDR) are restricted to the midrostrocaudal part of the raphe, and many of its serotonergic neurons are located in the adjacent ventrolateral periaqueductal gray (vlPAG) [3, 5, 8, 94]. As in the DRD, cell bodies in the lateral wings are smaller than those found in the rest of the DRN, and the dendritic arbor extends widely to all directions [6]. This subregion also shows a large population of GABAergic interneurons which appear to be important in the regulation of the activity of serotonergic neurons [6, 17, 95–97], as well as non-serotonergic, nitric oxide-producing neurons which project to the sensory thalamus [98, 99]. Non-serotonergic neurons from the lwDR show the higher action potential amplitude of all non-serotonergic neurons of other subregions, and they usually express 5-HT_{1A} receptors and respond with considerable hyperpolarization to the non-selective agonist 5-CT [6].

The lwDR receives dense projections from regions involved in autonomic regulation, including the lateral parabrachial nucleus, the medial portion of the nucleus of the solitary tract, and the A2 and C1 adrenergic cell groups [65, 100]. Important afferents also come from limbic structures, including the bed nucleus of the stria terminalis, the medial portion of the lateral habenula, the central amygdaloid nucleus, median preoptic nucleus, and lateral hypothalamus [40, 65, 101–108]. Efferents from the lwDR include the nucleus raphe magnus, lateral hypothalamus, rostroventrolateral medulla (RVLM), and periaqueductal gray (dlPAG) [109–112]. Importantly, the lwDR also projects to the ventral subnucleus of the DRN (DRV) [113]. Lesions of the lwDR increase tryptophan hydroxylase mRNA in DRV neurons [114], suggesting that the DRV is under inhibitory tone from the lwDR [3].

The origin of serotonin in both dorsal and ventral PAG is the nearby lwDR [111, 115]. Serotonin release in the dlPAG is controlled by glutamatergic mechanisms, as the infusion of NMDA decreases extracellular 5-HT concentrations in rats and elicits defense reactions, while stimulation of group II metabotropic glutamate receptors increase serotonin release; the latter effect is antagonized by the $GABA_AR$ antagonist bicuculline, suggesting an effect on GABAergic interneurons [116]. Morphological evidence has been gathered that about three-fourths of 5-HT-positive fibers in the dlPAG do not make synaptic contact with PAG neurons, strongly suggesting that the actions of serotonin in this subregion are predominantly through volume rather than wiring transmission [117].

Serotonergic neurons in the lwDR show faster in vitro firing rates, and have active and passive membrane properties that make them more excitable than DRV neurons, and their response to 5-CT is of higher magnitude than that observed in serotonergic neurons of the other subregions [6, 118]. These properties show a caudorostral gradient, with neurons in the rostral position of the lwDR showing the highest excitability [118]. An inverse gradient is observed in response to $5\text{-}HT_{1A}R$ agonists [118], and $5\text{-}HT_{1A}R$ antagonists produce the highest c-Fos-like immunoreactivity in this region [111]. The rostrocaudal gradient of $5\text{-}HT_{1A}R$ response is accompanied by a gradient of CRFergic synapses [119]. CRF release in this region is induced by swim stress, activating CRF_1 receptors in GABAergic neurons and ultimately inhibiting serotonin release [96].

Exposure to an open-field arena, social, swim, and interoceptive stress, as well as the administration of anxiogenic drugs, inescapable noxious cutaneous manipulation and induction of tonic immobility, activate neurons in the lwDR [38, 48, 94, 96, 120, 121–127]. Intravenous lactate infusions, which induce panic-like responses in panic-susceptive rats but not control animals [128], activate lwDR cells in control, but not in panic-prone animals, suggesting that this region is involved in "normative" responses to innocuous stressors [94, 122]. Restraint stress or systemic capsaicin injections increase c-Fos-like immunoreactivity as well as NADPH diaphorase activity in the lwDR, but these reactivities do not overlap [129, 130]. Neurons with c-Fos-like immunoreactivity are possibly GABAergic, since swim stress induces c-Fos in GABAergic cells of the lwDR [96]. It has been proposed that nitric oxide release in the lwDR during inescapable stress activates local GABAergic interneurons, reducing 5-HT release in the dlPAG [99]. This mechanism could be responsible for the observation that inescapable shocks potentiate subsequent fear conditioning and interfere with escape learning ("learned helplessness"), while escapable shock do not [131].

In the central nervous system, the serotonergic phenotype is controlled by an important series of transcription factors which are under the control of Nkx2.2 [63] and Mash1 [132]. Among those, plasmacytoma-expressed transcript 1 (Pet-1, also known as Fifth Ewing Variant, FEV), an E-twenty-six (ETS) transcription factor, is thought to be important in shaping dorsal raphe circuits and maintaining serotonergic homeostasis in the central nervous system [63, 133–136]. The expression of this transcription factor is restricted to serotonergic neurons; Pet-1 expression is under the transcriptional control of Sim1 (single-minded homolog 1), and *sim1*$^{-/-}$ mice show decreased 5-HT-like immunoreactivity that is restricted to the dorsal raphe nucleus and no *pet1* mRNA expression in the lateral wings of the raphe [137]. In the absence of Pet-1, a dramatic (~80 %) reduction is observed in the expression of tryptophan hydroxylase 2 [134, 135]; in 5-HT-like immunoreactivity in raphe nuclei [133, 138]; in serotonin content in the cortex, hippocampus and caudate nucleus [134]; in the expression of 5-HT_{1A} and 5-HT_{1B} autoreceptors in the DRN [133]; and in density of serotonin transporters and vesicular monoamine transporter isoform 2 (Vmat2) in raphe neurons [135]. The remaining serotonergic neurons show increased spontaneous firing of action potentials and decreased responsiveness to 8-OH-DPAT [133], and are actually increased in the lateral wings [135]. Importantly, $pet1^{-/-}$ mice conserve serotonergic innervation of the basolateral amygdala, habenula, and paraventricular hypothalamus (but not hippocampus or anterior cingulate cortex), as well as vGlut3 immunoreactivity [139], showing increased anxiety-like behavior when animals are tested in an elevated plus-maze after being exposed to three trials of a resident-intruder paradigm [134] but decreased anxiety-like behavior when tested in the EPM or elevated zero-maze without prior stress [136, 139]. Pharmacogenetic deletion of *pet1* in adult age abolishes expression of this gene in the DRN and produces an anxiogenic-like effect in the EPM without prior stress, associated with decreased expression of Tph2, SERT and vGlut3 [133]. Interestingly, $pet1^{-/-}$ mice show increased freezing to context and increased fear to tone in delayed and trace conditioning paradigms [39].

The transcription factors which regulate induction and maintenance of the remaining neurons in Pet-1 null mice, however, are unknown. It has been proposed that Pet-1 specifies the serotonergic phenotype of neurons in the DRN in parallel with an unknown, Pet-1-independent pathway [140]. The role of this Pet-1-independent pathway is dependent on prior stress, as embryonic deletion is anxiogenic in socially stressed mice but anxiolytic and panicogenic in non-stressed animals [134, 136, 139]. Pet-1 seems to be important in regulating the distribution of dorsal raphe neurons, as 5-HT-like immunoreactivity is increased in the lateral wings of the DRN in knockout mice [135]. Sim1 emerges as a regulator of this distribution, since $sim1^{-/-}$ animals do not express Pet-1 mRNA in the lwDR at birth [137]. Thus, an Sim1 →Pet-1 pathway is able to regulate the stress-responsive, fear-regulating cells of the lateral wings of the PAG. A problem with this approach, of course, is that $sim1^{-/-}$ animals are not viable, dying after birth due to hypothalamic defects [141], and therefore their behavioral phenotype has not been analyzed.

Other possible transcription factors are Gata2 and Gata3, zinc-finger transcription factors that regulate the serotonergic phenotype independent of Pet-1 [142]. In immortalized mesencephalic MN9D cells, *sim1* knockdown decreases *gata2* mRNA by about 30 % [137]. In Gata3 knockout animals, Tph2 mRNA expression is decreased in the dorsal raphe, especially at the lwDR [133]; the expression of the vesicular monoamine transporter 2 (Vmat2), SERT, and aromatic amino acid decarboxylase (AADC) is also decreased, as is 5-HT and 5-HIAA content in the forebrain and spinal cord. Nonetheless, the expression of 5-HT_{1A} autoreceptors, as well as their function, is conserved in these animals [133].

References

1. Silveira MCL, Zangrossi H Jr, Viana MB, Silveira R, Graeff FG (2001) Differential expression of Fos protein in the rat brain induced by performance of avoidance or escape in the elevated T-maze. Behav Brain Res 126:13–21
2. Lowry CA, Hale MW (2010) Serotonin and the neurobiology of anxious states. In: Muller CP, Jacobs BL (eds) The behavioral neurobiology of serotonin. Academic Press, London, pp 379–397
3. Hale MW, Lowry CA (2011) Functional topography of midbrain and pontine serotonergic systems: implications for synaptic regulation of serotonergic circuits. Psychopharmacology 213:243–264
4. Baker KG, Halliday GM, Tork I (1990) Cytoarchitecture of the human dorsal raphe nucleus. J Comp Neurol 301:147–161
5. Abrams JK, Johnson PL, Hollis JH, Lowry CA (2004) Anatomic and functional topography of the dorsal raphe nucleus. Ann N Y Acad Sci 1018:46–57
6. Calizo LH, Ma X, Pan Y, Lemos J, Craige C, Heemstra L, Beck SG (2011) Raphe serotonin neurons are not homogenous: electrophysiological, morphological and neurochemical evidence. Neuropsychopharmacology 61:524–543
7. Gasser PJ, Orchinik M, Raju I, Lowry CA (2009) Distribution of organic cation transporter 3, a corticosterone-sensitive monoamine transporter, in the rat brain. J Comp Neurol 512:529–555
8. Clark MS, McDevitt RA, Neumaier JF (2006) Quantitative mapping of tryptophan hydroxylase-2, 5-HT$_{1A}$, 5-HT$_{1B}$, and serotonin transporter expression across the anteroposterior axis of the rat dorsal and median raphe nuclei. J Comp Neurol 498:611–623
9. Dahlin A, Xia L, Kong W, Hevner R, Wang J (2007) Expression and immunolocalization of the plasma membrane monoamine transporter in the brain. Neuroscience 146:1193–1211
10. Stockmeier CA, Shapiro LA, Haycock JW, Thompson PA, Lowy MT (1996) Quantitative subregional distribution of serotonin$_{1A}$ receptors and serotonin transporters in the human dorsal raphe. Brain Res 727:1–12
11. Commons KG, Connolley KR, Valentino RJ (2003) A neurochemically distinct dorsal raphe-limbic circuit with a potential role in affective disorders. Neuropsychopharmacology 28:205–215
12. Rainnie DG, Fernhout BJH, Shinnick-Gallagher P (1992) Differential actions of corticotropin releasing factor on basolateral and central amygdaloid neurons, in vitro. J Pharmacol Exp Ther 263:846–858
13. Kirby LG, Rice KC, Valentino RJ (2000) Effects of corticotropin-releasing factor on neuronal activity in the serotonergic dorsal raphe nucleus. Neuropsychopharmacology 22:148–162
14. Tsetsenis T, Ma XH, Lo Iacono L, Beck SG, Gross C (2007) Suppression of conditioning to ambiguous cues by pharmacogenetic inhibition of the dentate gyrus. Nat Neurosci 10:896–902
15. Gozzi A, Jain A, Giovanelli A, Bertollini C, Crestan V, Schwarz AJ, Tsetsenis T, Ragozzino D, Gross CT, Bifone A (2010) A neural switch for active and passive fear. Neuron 67:656–666
16. Valentino RJ, Lucki I, Van Bockstaele EJ (2010) Corticotropin-releasing factor in the dorsal raphe nucleus: linking stress coping and addiction. Brain Res 1314:29–37
17. Day HEW, Greenwood BN, Hammack SE, Watkins LR, Fleshner M, Maier SF, Campeau S (2004) Differential expression of 5-HT-1A, a_{1b} adrenergic, CRF-R1, and CRF-R2 receptor mRNA in serotonergic, g-aminobutyric acidergic, and catecholaminergic cells of the rat dorsal raphe nucleus. J Comp Neurol 474:364–378
18. Pernar L, Curtis AL, Vale WW, Rivier JE, Valentino RJ (2004) Selective activation of corticotropin-releasing factor-2 receptors on neurochemically identified neurons in the rat dorsal raphe nucleus reveals dual actions. J Neurosci 24:1311–1350

19. McEuen JG, Beck SG, Bale TL (2008) Failure to mount adaptive responses to stress results in dysregulation and cell death in the midbrain raphe. J Neurosci 28:8169–8177
20. Häring M, Marsicano G, Lutz B, Monory K (2007) Identification of the cannabinoid receptor type 1 in serotonergic cells of raphe nuclei in mice. Neuroscience 146:1212–1219
21. Marin WJ, Patrick SL, Coffin PO, Tsou K, Walker JM (1995) An examination of the central sites of action of cannabinoid-induced antinociception in the rat. Life Sci 56:2103–2109
22. Mendiguren A, Pineda J (2009) Effect of the CB_1 receptor antagonists rimonabant and AM251 on the firing rate of dorsal raphe nucleus neurons in rat brain slices. Br J Pharmacol 158:1579–1587
23. Haj-Dahmane S, Shen R-Y (2009) Endocannabinoids suppress excitatory synaptic transmission to dorsal raphe serotonin neurons through the activation of presynaptic CB_1 receptors. J Pharmacol Exp Ther 331:186–196
24. Tzavara ET, Davis RJ, Perry KW, Li X, Salhoff C, Bymaster FP, Witkin JM, Nomikos GG (2003) The CB1 receptor antagonist SR141716A selectively increases monoaminergic neurotransmission in the medial prefrontal cortex: implications for therapeutic actions. Br J Pharmacol 138:544–553
25. Aso E, Renoir T, Mengod G, Ledent C, Hamon M, Maldonado R, Lanfumey L, Valverde O (2009) Lack of CB_1 receptor activity impairs serotoninergic negative feedback. J Neurochem 109:935–944
26. Egashira N, Mishima K, Katsurabayashi S, Yoshitake T, Matsumoto Y, Ishida J, Yamaguchi M, Iwasaki K, Fujiwara M (2002) Involvement of 5-hydroxytryptamine neuronal system in delta(9)-tetrahydrocannabinol-induced impairment of spatial memory. Eur J Pharmacol 445:221–229
27. Darmani NA, Janoyan JJ, Kumar N, Crim JL (2003) Behaviorally active doses of the CB_1 receptor antagonist SR 14176A increase brain serotonin and dopamine levels and turnover. Pharmacol Biochem Behav 75:777–787
28. Thiemann G, Watt CA, Ledent C, Molleman A, Hasenöhrl RU (2009) Modulation of anxiety by acute blockade and genetic deletion of the CB_1 cannabinoid receptor in mice together with biogenic amine changes in the forebrain. Behav Brain Res 200:60–67
29. Marco EM, Perez-Alvarez L, Borcel E, Rubio M, Guaza C, Ambrosio E, File SE, Viveros MP (2004) Involvement of 5-HT1A receptors in behavioural effects of the cannabinoid receptor agonist CP 55,940 in male rats. Behav Pharmacol 15:21–27
30. Imai H, Steindler D, Kitai ST (1986) The organization of divergent axonal projections from the midbrain raphe nuclei in the rat. J Comp Neurol 243:363–380
31. Lowry CA, Jonhson PL, Hay-Schmidt A, Mikkelsen J, Shekhar A (2005) Modulation of anxiety circuits by serotonergic systems. Stress 8:233–246
32. Li Y-Q, Jai H-G, Rao Z-R, Shi J-W (1990) Serotonin-, substance P- or leucine-enkephalin-containing neurons in the midbrain periaqueductal gray and nucleus raphe dorsalis send projection fibers to the central amygdaloid nucleus in the rat. Neurosci Lett 120:124–127
33. Steinbusch HW, Nieuwenhuys R, Verhofstad AA, van der Kooy D (1981) The nucleus raphe dorsalis of the rat and its projection upon the caudatoputamen. A combined cytoarchitectonic, immunohistochemical and retrograde transport study. J Physiol 77:157–174
34. Steinbusch HW, van der Kooy D, Verhofstad AA, Pellegrino A (1980) Serotonergic and non-serotonergic projections from the nucleus raphe dorsalis to the caudate-putamen complex in the rat, studied by a combined immunofluorescence and fluorescent retrograde axonal labeling technique. Neurosci Lett 19:137–142
35. Köhler C, Steinbusch HW (1982) Identification of serotonin and non-serotonin-containing neurons of the mid-brain raphe projecting to the entorhinal area and the hippocampal formation. A combined immunohistochemical and fluorescent retrograde tracing study in the rat brain. Neuroscience 7:951–975
36. Canteras NS, Shammah-Lagnado SJ, Silva BA, Ricardo JA (1990) Afferent connections of the subthalamic nucleus: a combined retrograde and anterograde horseradish peroxidase study in the rat. Brain Res 513:43–59

37. Waterhouse BD, Mihailoff GA, Baack JC, Woodward DJ (1986) Topographical distribution of dorsal and median raphe neurons projecting to motor, sensorimotor, and visual cortical areas in the rat. J Comp Neurol 249:460–481
38. Abrams JK, Johnson PL, Hay-Schmidt A, Mikkelsen J, Shekhar A, Lowry CA (2005) Serotonergic systems associated with arousal and vigilance behaviors following administration of anxiogenic drugs. Neuroscience 133:983–997
39. van Bockstaele EJ, Biswas A, Pickel VM (1993) Topography of serotonin neurons in the dorsal raphe nucleus that send axon collaterals to the rat prefrontal cortex and nucleus accumbens. Brain Res 624:188–198
40. Rizvi TA, Ennis M, Behbehani MM, Shipley MT (1991) Connections between the central nucleus of the amygdala and the midbrain periaqueductal gray: topography and reciprocity. J Comp Neurol 303:121–131
41. Weller KL, Smith DA (1982) Afferent connections to the bed nucleus of the stria terminalis. Brain Res 232:255–270
42. Petrov T, Krukoff TL, Jhamandas JH (1994) Chemically defined collateral projections from the pons to the central nucleus of the amygdala and paraventricular nucleus in the rat. Cell Tissue Res 277:289–295
43. Li Y-Q, Takada M, Matsuzaki S, Shinonaga Y, Mizuno N (1993) Identification of periaqueductal gray and dorsal raphe nucleus neurons projecting to both the trigeminal sensory complex and forebrain structures: a fluorescent retrograde double-labeling study in the rat. Brain Res 623:267–277
44. Chen S, Su HS (1990) Afferent connections of the thalamic paraventricular and parataenial nuclei in the rat—a retrograde tracing study with iontophoretic application of fluoro-gold. Brain Res 522:1–6
45. Cornwall J, Phillipson OT (1988) Afferent projections to the dorsal thalamus of the rat as shown by retrograde lectin transport. II. The midline nuclei. Brain Res Bull 21:147–161
46. Jasmin L, Burkey AR, Granato A, Ohara PT (2004) Rostral agranular insular cortex and pain areas of the central nervous system: a tract-tracing study in the rat. J Comp Neurol 468:425–440
47. Hioki H, Nakamura H, Ma Y-F, Konno M, Hakayama T, Nakamura KC, Fujiyama F, Kaneko T (2009) Vesicular glutamate transporter 3-expressing nonserotonergic projection neurons constitute a subregion in the rat midbrain raphe nuclei. J Comp Neurol 518:668–686
48. Hale MW, Hay-Schmidt A, Mikkelsen J, Poulsen B, Bouwknecht JA, Evans AK, Stamper CE, Shekhar A, Lowry CA (2008) Exposure to an open-field arena increases c-Fos expression in a subpopulation of neurons in the dorsal raphe nucleus, including neurons projecting to the basolateral amygdaloid complex. Neuroscience 157:733–748
49. Kirby LG, Pernar L, Valentino RJ, Beck SG (2003) Distinguishing characteristics of serotonin and non-serotonin-containing cells in the dorsal raphe nucleus: electrophysiological and immunohistochemical studies. Neuroscience 116:669–683
50. Wang Q-P, Guan J-L, Nakai Y (1995) Distribution and synaptic relations of NOS neurons in the dorsal raphe nucleus: a comparison to 5-HT neurons. Brain Res Bull 37:177–187
51. Grahn RE, Watkins LR, Maier SF (2000) Impaired escape performance and enhanced conditioned fear in rats following exposure to an uncontrollable stressor are mediated by glutamate and nitric oxide in the dorsal raphe nucleus. Behav Brain Res 112:33–41
52. Beijamini V, Guimarães FS (2006) Activation of neurons containing the enzyme nitric oxide synthase following exposure to an elevated plus maze. Brain Res Bull 69:347–355
53. Beijamini V, Guimarães FS (2006) c-Fos expression increase in NADPH-diaphorase positive neurons after exposure to a live cat. Behav Brain Res 170:52–61
54. Xu Z-QD, Hökfelt T (1997) Expression of galanin and nitric oxide synthase in subpopulations of serotonin neurons of the rat dorsal raphe nucleus. J Chem Neuroanat 13:169–187
55. Hale MW, Johnson PL, Westerman AM, Abrams JK, Shekhar A, Lowry CA (2010) Multiple anxiogenic drugs recruit a parvalbumin-containing subpopulation of GABAergic

interneurons in the basolateral amygdala. Prog Neuropsychopharmacol Biol Psychiatry 34:1285–1293
56. Spannuth BM, Hale MW, Evans AK, Lukkes JL, Campeau S, Lowry CA (2011) Investigation of a central nucleus of the amygdala/dorsal raphe nucleus serotonergic circuit implicated in fear-potentiated startle. Neuroscience 179:104–119
57. Maier SF, Watkins LR (2005) Stressor controllability and learned helplessness: the roles of the dorsal raphe nucleus, serotonin, and corticotropin-releasing factor. Neurosci Biobehav Rev 29:829–841
58. Vanderhaeghen JJ, Lotstra F, De MJ, Gilles C (1980) Immunohistochemical localization of cholekystokinin- and gastrin-like peptides in the brain and hypophysis of the rat. Proc Natl Acad Sci U S A 77:1190–1194
59. Smith GST, Savery D, Marden C, Lopez Costa JJ, Averill S, Priestley JV, Rattray M (1994) Distribution of messenger RNAs encoding enkephalin, substance P, somatostatin, galanin, vasoactive intestinal polypeptide, neuropeptide Y, and calcitonin gene-related peptide in the midbrain periaqueductal grey in the rat. J Comp Neurol 350:23–40
60. Sutin EL, Jacobowitz DM (1988) Immunocytochemical localization of peptides and other neurochemicals in the rat laterodorsal tegmental nucleus and adjacent area. J Comp Neurol 270:243–270
61. Fu W, Le Maître E, Fabre V, Bernard J-F, Xu Z-QD, Hökfelt T (2010) Chemical neuroanatomy of the dorsal raphe nucleus and adjacent structures of the mouse brain. J Comp Neurol 518:3464–3494
62. Soiza-Reilly M, Commons KG (2011) Glutamatergic drive of the dorsal raphe nucleus. J Chem Neuroanat 41:247–255
63. Cheng L, Chen C-L, Luo P, Tan M, Qiu M, Johnson R, Ma Q (2003) Lmx1b, Pet-1, and Nkx2.2 coordinately specify serotonergic neurotransmitter phenotype. J Neurosci 23:9961–9967
64. Amilhon B, Lepicard È, Renoir T, Mongeau R, Popa D, Poirel O, Miot S, Gras C, Gardier AM, Gallego J, Hamon M, Lanfumey L, Gasnier B, Giros B, El Mestikawy S (2010) VGLUT3 (vesicular glutamate transporter type 3) contribution to the regulation of serotonergic transmission and anxiety. J Neurosci 30:2198–2210
65. Lee HS, Kim M-A, Valentino RJ, Waterhouse BD (2003) Glutamatergic afferent projections to the dorsal raphe nucleus of the rat. Brain Res 963:57–71
66. Lind RW (1986) Bi-directional, chemically specified neural connections between the subfornical organ and the midbrain raphe system. Brain Res 384:250–261
67. Mikkelsen JD, Hay-Schmidt A, Larsen PJ (1997) Central innervation of the rat ependyma and subcommissural organ with special reference to ascending serotoninergic projections from the raphe nuclei. J Comp Neurol 384:556–568
68. Simpson KL, Fisher TM, Waterhouse BD, Lin RC (1998) Projection patterns from the raphe nuclear complex to the ependymal wall of the ventricular system in the rat. J Comp Neurol 399:61–72
69. Kocsis B, Varga V, Dahan L, Sik A (2006) Serotonergic neuron diversity: identification of raphe neurons with discharges time-locked to the hippocampal theta rhythm. Proc Natl Acad Sci U S A 103:1059–1064
70. Aghajanian GK (1990) Use of brain slices in the study of serotoninergic pacemaker neurons of the brainstem raphe nuclei. Wiley, New York
71. Staub DR, Evans AK, Lowry CA (2006) Evidence supporting a role for corticotropin-releasing factor type 2 (CRF_2) receptors in the regulation of subpopulations of serotonergic neurons. Brain Res 1070:77–89
72. Staub DR, Spiga F, Lowry CA (2005) Urocortin 2 increases c-Fos expression in topographically organized subpopulations of serotonergic neurons in the rat dorsal raphe nucleus. Brain Res 1044:176–189
73. Hale MW, Stamper CE, Staub DR, Lowry CA (2010) Urocortin 2 increases c-Fos expression in serotonergic neurons projecting to the ventricular/periventricular system. Exp Neurol 224:271–281

74. Petrov T, Krukoff TL, Jhamandas JH (1992) The hypothalamic paraventricular and lateral parabrachial nuclei receive collaterals from raphe nucleus neurons: a combined double retrograde and immunocytochemical study. J Comp Neurol 318:18–26
75. Petrov T, Krukoff TL, Jhamandas JH (1994) Chemically defined collateral projections from the pons to the central nucleus of the amygdala and hypothalamic paraventricular nucleus in the rat. Cell Tissue Res 277:289–295
76. Lowry CA (2002) Functional subsets of serotonergic neurones: implications for control of the hypothalamic-pituitary-adrenal axis. J Neuroendocrinol 14:911–923
77. Singewald N, Sharp T (2000) Neuroanatomical targets of anxiogenic drugs in the hindbrain as revealed by Fos immunocytochemistry. Neuroscience 98:759–770
78. Evans AK, Heerkens JLT, Lowry CA (2009) Acoustic stimulation in vivo and corticotropin-releasing factor in vitro increase tryptophan hydroxylase activity in the rat caudal dorsal raphe nucleus. Neurosci Lett 455:36–41
79. Evans AK, Reinders N, Ashford KA, Christie IN, Wakerley JB, Lowry CA (2008) Evidence for serotonin synthesis-dependent regulation of in vitro neuronal firing rates in the midbrain raphe complex. Eur J Pharmacol 590:136–149
80. Grahn RE, Will MJ, Hammack SE, Maswood S, McQueen MB, Watkins KC, Maier SF (1999) Activation of serotonin-immunoreactive cells in the dorsal raphe nucleus in rats exposed to an uncontrollable stressor. Brain Res 826:35–43
81. Maswood S, Barter JE, Watkins LR, Maier SF (1998) Exposure to inescapable but not escapable shock increases extracellular levels of 5-HT in the dorsal raphe nucleus of the rat. Brain Res 783:115–120
82. McDevitt RA, Hiroi R, Mackenzie SM, Robin NC, Cohn A, Kim JJ, Neumaier JF (2011) Serotonin 1B autoreceptors originating in the caudal dorsal raphe nucleus reduce expression of fear and depression-like behavior. Biol Psychiatry 69:780–787
83. Clark MS, Sexton TJ, McClain M, Root DC, Kohen R, Neumaier JF (2002) Overexpression of 5-HT1B receptor in dorsal raphe nucleus using herpes simplex virus gene transfer increases anxiety behavior after inescapable stress. J Neurosci 22:4550–4562
84. Hammack SE, Richey KJ, Schmid MJ, LoPresti ML, Watkins LR, Maier SF (2002) The role of corticotropin-releasing hormone in the dorsal raphe nucleus in mediating the behavioral consequences of uncontrollable stress. J Neurosci 22:1020–1026
85. Hammack SE, Schmid MJ, LoPresti ML, Der-Avakian A, Pellymounter MA, Foster AC, Watkins LR, Maier SF (2003) Corticotropin releasing hormone type 2 receptors in the dorsal raphe nucleus mediate the behavioral consequences of uncontrollable stress. J Neurosci 23:1019–1025
86. Pernar L, Curtis AL, Vale WW, Rivier JE, Valentino RJ (2004) Selective activation of corticotropin-releasing factor-2 receptors on neurochemically identified neurons in the rat dorsal raphe nucleus reveals dual actions. J Neurosci 24:1305–1311
87. Waselus M, Valentino RJ, Van Bockastaele EJ (2011) Collateralized dorsal raphe nucleus projections: a mechanism for the integration of diverse functions during stress. J Chem Neuroanat 41:266–280
88. Grahn RE, Watkins LR, Maier SF (2000) Impaired escaped performance and enhanced conditioned fear in rats following exposure to an uncontrollable stressor are mediated by glutamate and nitric oxide in the dorsal raphe nucleus. Behav Brain Res 112:33–41
89. Ferraro G, Montalbano ME, Sardo P, La Grutta V (1996) Lateral habenular influence on dorsal raphe neurons. Brain Res Bull 41:47–52
90. Amat J, Sparks PD, Matus-Amat P, Griggs J, Watkins LR, Maier SF (2001) The role of the habenular complex in the elevation of dorsal raphe nucleus serotonin and the changes in the behavioral responses produced by uncontrollable stress. Brain Res 97:118–126
91. Neckers LM, Schwartz JP, Wyatt RJ, Speciale SG (1979) Substance P afferents from the habenula innervate the dorsal raphe nucleus. Exp Brain Res 37:619–623
92. Lee A, Mathuru AS, Teh C, Kibat C, Korzh V, Penney TB, Jesuthasan S (2010) The habenula prevents helpless behavior in larval zebrafish. Curr Biol 20:2211–2216

93. Amat J, Baratta MV, Paul E, Bland ST, Watkins LR, Maier SF (2005) Medial prefrontal cortex determines how stressor controllability affects behavior and dorsal raphe nucleus. Nat Neurosci 8:365–371
94. Johnson PL, Lightman SL, Lowry CA (2004) A functional subset of serotonergic neurons in the rat ventrolateral periaqueductal gray implicated in the inhibition of serotonergic neurons in the rat ventrolateral periaqueductal gray implicated in the inhibition of sympathoexcitation and panic. Ann N Y Acad Sci 1018:58–64
95. Jolas T, Aghajanian GK (1997) Opioids suppress spontaneous and NMDA-induced inhibitory postsynaptic currents in the dorsal raphe nucleus of the rat in vitro. Brain Res 755:229–245
96. Roche M, Commons KG, Peoples A, Valentino RJ (2003) Circuitry underlying regulation of the serotonergic system by swim stress. J Neurosci 23:970–977
97. Boothman LJ, Sharp T (2005) A role for midbrain raphe g-aminobutyric acid neurons in 5-hydroxytryptamine feedback control. NeuroReport 16:891–896
98. Simpson KL, Waterhouse BD, Lin RCS (2003) Differential expression of nitric oxide in serotonergic projection neurons: neurochemical identification of dorsal raphe inputs to rodent trigeminal somatosensory targets. J Comp Neurol 466:495–512
99. Vasudeva RK, Lin RCS, Simpson KL, Waterhouse BD (2011) Functional organization of the dorsal raphe efferent system with special consideration of nitrergic cell groups. J Chem Neuroanat 41:281–293
100. Herbert H (1992) Evidence for projections from medullary nuclei onto serotonergic and dopaminergic neurons in the midbrain dorsal raphe nucleus of the rat. Cell Tissue Res 270:149–156
101. Kim U (2009) Topographic commissural and descending projections of the habenula in the rat. J Comp Neurol 513:173–187
102. Gray TS, Magnuson DJ (1987) Galanin-like immunoreactivity within amygdaloid and hypothalamic neurons that project to the midbrain central grey in rat. Neurosci Lett 83:264–268
103. Gray TS, Magnuson DJ (1992) Peptide immunoreactive neurons in the amygdala and the bed nucleus of the stria terminalis project to the midbrain central gray in the rat. Peptides 13:451–460
104. Luiten PG, ter Horst GJ, Steffens AB (1987) The hypothalamus, intrinsic connections and outflow pathways to the endocrine system in relation to the control of feeding and metabolism. Prog Neurobiol 28:1–54
105. Saper CB, Swanson LW, Cowan WM (1979) An autoradiographic study of the efferent connections of the lateral hypothalamic area in the rat. J Comp Neurol 183:689–706
106. Holstege G, Meiners L, Tan K (1985) Projections of the bed nucleus of the stria terminalis to the mesencephalon, pons, and medulla oblongata in the cat. Exp Brain Res 58:379–391
107. Simerly RB, Swanson LW (1988) Projections of the medial preoptic nucleus: a *Phaseolus vulgaris* leucoagglutinin anterograde tract-tracing study in the rat. J Comp Neurol 270:209–242
108. Swanson LW, Mogenson GJ, Simerly RB, Wu M (1987) Anatomical and electrophysiological evidence for a projection from the medial preoptic area to the 'mesencephalic and subthalamic regions' in the rat. Brain Res 405:108–122
109. Ljubic-Thibal V, Morin A, Diksic M, Hamel E (1999) Origin of the serotonergic innervation to the rat dorsolateral hypothalamus: retrograde transport of cholera toxin and upregulation of tryptophan hydroxylase mRNA expression following selective nerve terminals lesion. Synapse 32:177–186
110. Stezhka VV, Lovick TA (1997) Projections from dorsal raphe nucleus to the periaqueductal grey matter: studies in slices of rat midbrain maintained in vitro. Neurosci Lett 230:57–60
111. Beitz AJ (1982) The sites of origin of brain stem neurotensin and serotonin projections to the rodent nucleus raphe magnus. J Neurosci 2:829–842
112. Bago M, Marson L, Dean C (2002) Serotonergic projections to the rostroventrolateral medulla from midbrain and raphe nuclei. Brain Res 945:249–258

113. Peyron C, Petit J-M, Rampon C, Jouvet M, Luppi P-H (1998) Forebrain afferents to the rat dorsal raphe nucleus demonstrated by retrograde and anterograde tracing methods. Neuroscience 82:443–468
114. Bendotti C, Servadio A, Forloni G, Angeretti N, Samanin R (1990) Increased tryptophan hydroxylase mRNA in raphe serotonergic neurons spared by 5,7-dihydroxytryptamine. Mol Brain Res 8:343–348
115. Kiwat GC, Basbaum AI (1990) Organization of tyrosine hydroxylase- and serotonin-immunoreactive brainstem neurons with axon collaterals to the periaqueductal gray and the spinal cord in the rat. Brain Res 24:83–94
116. Maione S, Palazzo E, De Novellis V, Stella L, Leyva J, Rossi F (1998) Metabotropic glutamate receptors modulate serotonin release in the rat periaqueductal gray matter. Naunyn-Schmiedeberg's Arch Pharmacol 358:411–417
117. Lovick TA, Parry DM, Stezhka VV, Lumb BM (1999) Serotonergic transmission in the periaqueductal gray matter in relation to aversive behavior: morphological evidence for direct modulatory effects on identified output neurons. Neuroscience 95:763–772
118. Crawford LK, Craige CP, Beck SG (2010) Increased intrinsic excitability of lateral wing serotonin neurons of the dorsal raphe: a mechanism for selective activation in stress circuits. J Neurophysiol 103:2652–2663
119. Valentino RJ, Liouterman L, van Bockastaele EJ (2001) Evidence for regional heterogeneity in corticotropin-releasing factor interactions in the dorsal raphe nucleus. J Comp Neurol 435:450–463
120. Commons KG (2008) Evidence for topographically organized endogenous 5-HT-1A receptor-dependent feedback inhibition of the ascending serotonin system. Eur J Neurosci 27:2611–2618
121. Gardner KL, Thrivikraman KV, Lightman SL, Plotsky PM, Lowry CA (2005) Early life experience alters behavior during social defeat: focus on serotonergic systems. Neuroscience 136:181–191
122. Johnson PL, Lowry CA, Truitt W, Shekhar A (2008) Disruption of GABAergic tone in the dorsomedial hypothalamus attenuates responses in a subset of serotonergic neurons in the dorsal raphe nucleus following lactate-induced panic. J Psychopharmacol 22:642–652
123. Berton O, Covington HE 3rd, Ebner L, Tsankova NM, Carle TL, Ulery P, Bhonsle A, Barrot M, Krishnan V, Singewald GM, Singewald N, Birnbaum S, Neve rL, Nestler EJ (2007) Induction of DFosB in the periaqueductal gray by stress promotes active coping responses. Neuron 55:289–300
124. Vieira EB, Menescal-de-Oliveira L, Leite-Panissi CRA (2011) Functional mapping of the periaqueductal gray matter involved in organizing tonic immobility behavior in guinea pigs. Behav Brain Res 216:94–99
125. Keay KA, Clement CI, Depaulis A, Bandler R (2001) Different representations of inescapable noxious stimuli in the periaqueductal gray and upper cervical spinal cord of freely moving rats. Neurosci Lett 313:17–20
126. Martinez J, Phillips PJ, Herbert J (1998) Adaptation in patterns of *c-fos* expression in the brain associated with exposure to either single or repeated social stress in male rats. Eur J Neurosci 10:20–33
127. Chung KKK, Martinez J, Herbert J (2000) *c-fos* expression, behavioural, endocrine and autonomic responses to acute social stress in male rats after chronic restraint: modulation by serotonin. Neuroscience 95:453–463
128. Johnson PL, Truitt WA, Fitz SD, Lowry CA, Shekhar A (2008) Neural pathways underlying lactate-induced panic. Neuropsychopharmacology 33:2093–2107
129. Okere CO, Waterhouse BD (2006) Activity-dependent heterogeneous populations of nitric oxide synthase neurons in the rat dorsal raphe nucleus. Brain Res 1086:117–132
130. Okere CO, Waterhouse BD (2006) Acute restraint increases NADPH-diaphorase staining in distinct subregions of the rat dorsal raphe nucleus: implications for raphe serotonergic and nitrergic transmission. Brain Res 1119:174–181

131. Maier SF (1990) Role of fear in mediating shuttle escape learning deficit produced by inescapable shock. J Exp Psychol: Anim Behav Process 16:137–149
132. Pattyn A, Simplicio N, van Doorninck JH, Goridis C, Guillemot F, Brunet J-F (2004) *Ascl1/Mash1* is required for the development of central serotonergic neurons. Nat Neurosci 7:589–595
133. Liu C, Maejima T, Wyler SC, Casadesus G, Herlitze S, Deneris ES (2010) *Pet-1* is required across different stages of life to regulate serotonergic function. Nat Neurosci 13:1190–1198
134. Hendricks TJ, Fyodorov DV, Wegman LJ, Lelutiu NB, Pehek EA, Yamamoto B, Silver J, Weeber EJ, Sweatt JD, Deneris ES (2003) Pet-1 ETS gene plays a critical role in 5-HT neuron development and is required for normal anxiety-like and aggressive behavior. Neuron 37:233–247
135. Krueger KC, Deneris ES (2008) Serotonergic transcription of human *FEV* reveals direct GATA factor interactions and fate of Pet-1-deficient serotonin neuron precursors. J Neurosci 28:12748–12758
136. Schaefer TL, Vorhees CV, Williams MT (2009) Mouse plasmacytoma-expressed transcript 1 knock out induced 5-HT disruption results in a lack of cognitive deficits and an anxiety phenotype complicated by hypoactivity and defensiveness. Neuroscience 164:1431–1443
137. Osterberg N, Wiehle M, Oehlke O, Heidrich S, Xu C, Fan C-M, Krieglstein K, Roussa E (2011) Sim1 is a novel regulator in the differentiation of mouse dorsal raphe serotonergic neurons. Plos One 6:e19239
138. Scott MM, Krueger KC, Deneris ES (2005) A differentially autoregulated Pet-1 enhancer region is a critical target of the transcriptional cascade that governs serotonin neuron development. J Neurosci 25:2628–2636
139. Kiyasova V, Fernandez SP, Laine J, Stankovski L, Muzerelle A, Doly S, Gaspar P (2011) A genetically defined morphologically and functionally unique subset of 5-HT neurons in the mouse raphe nucleu. J Neurosci 31:2756–2768
140. Cordes SP (2005) Molecular genetics of the early development of hindbrain serotonergic neurons. Clin Genet 68:487–494
141. Michaud JL, Rosenquist T, May NR, Holdener BC, Fan C-M (1998) Development of neuroendocrine lineages requires the bHLH-PAS transcription factor SIM1. Genes Dev 12:3264–3275
142. Craven SE, Lim K-C, Ye W, Engel JD, de Sauvage F, Rosenthal A (2004) Gata2 specifies serotonergic neurons downstream of sonic hedgehog. Development 131:1165–1173

Chapter 6
General Conclusions

While the Deakin-Graeff hypothesis of serotonin function in anxiety disorders was appropriate to explain the data collected on the behavioral effects of serotonergic drugs microinjected in the amygdala or PAG, posterior analyses proved the reality to be far more nuanced than this hypothesis could cover. At the same time, while 5-HT_{1A} agonists decrease anxiety- and fear-like behavior when microinjected in the basolateral amygdala [1] and PAG [2, 3], they increase anxiety-like behavior when microinjected in the septo-hippocampal system [4–8]. Similarly, while 5-HT_2 receptor agonists in the PAG reduce measures of fear and anxiety [3, 9, 10], they increase anxiety-like behavior when microinjected into the basolateral amygdala [11, 12] and septo-hippocampal system [13]. Overall, these pharmacological results suggest that the regulation of anxiety-like behavior by serotonin is complex, with different receptors producing different results depending on the structure.

An important observation regarding the role of serotonin in anxiety is that, in spite of the different roles of serotonin in these structures, the origin of this serotonin is presumably the same—viz., the dorsal raphe nucleus. Nonetheless, the DRN is not homogeneous, with different subregions presenting different neurochemistry, electrophysiological properties, responsiveness to stress, and hodology. This topographic and topological organization can explain the initial observation of the Deakin-Graeff hypothesis that serotonin is "anxiogenic in the amygdala an anxiolytic in the PAG"; in conjunction with the pharmacological and anatomical studies on the microcircuit of the amygdaloid nuclei and periaqueductal gray, these observations might explain the discrepancy in the microinjection results. The discrepancy between the effects of the 5-HT_{1A} antagonist WAY 100635 on responses mediated by the ventrolateral vs. dorsolateral PAG, for example, is probably related to the increased sensitivity of 5-HT_{1A} receptors in the lwDR in relation to other DRN subdivisions [14, 15].

An important avenue in the neuroanatomical and pharmacological approach to the study of anxiety is the description of the sustained role of developmental genes in organizing the neurochemical and physiological properties of serotonergic neurons in the different subdivisions. For example, *sim1* has been shown to

C. Maximino, *Serotonin and Anxiety*, SpringerBriefs in Neuroscience,
DOI: 10.1007/978-1-4614-4048-2_6, © The Author(s) 2012

positively regulate the lwDR expression of the RGS4 [16], a protein which mediates 5-HT_{1A}R-dependent serotonin release [17]. While *pet1* is necessary postnatally for the expression of 5-HT_{1A} receptors and for the electrophysiological effects of 8-OH-DPAT on raphe slices, these effects were not observed when *pet1* was conditionally knocked out in adulthood in spite of an effect on anxiety-like behavior [18]. Since Sim1 knockout mice are not viable, the behavioral role of this transcription factor is not amenable for analysis in this species, and therefore the jury is still out on whether it indeed represents an important differentiator of the physiological and neurochemical properties of the lwDR.

While role of serotonin in mediating fear/panic responses is relatively defined, the same cannot be said about the serotonergic regulation of anxiety-like states. Thus, while forebrain 5-HT_{2C} receptors are clearly involved in the generation of these states [12, 13, 19–22], several questions remain regarding the role of 5-HT_{1A} receptors in these structures. While studies using genetically modified animals suggest that forebrain 5-HT_{1A} receptors inhibit anxiety [23], pharmacological [1, 4–8] and more refined genetic [24, 25] experiments suggest a far more complex relationship.

As our knowledge of the neuroanatomical, physiological, and pharmacological organization of the portions of the serotonergic system which modulate anxiety-like behavior is at present extensive, the future landscape is still very much open for inquiry. What is the relationship between the functional right lateralization of the behavioral inhibition and cerebral aversive systems and serotonin innervation? What is the true relationship between 5-HT_{2C} receptor editing and anxiety-like behavior? Are there other serotonergic "neural switches" in the CNS which have not been identified? These are exciting questions which will hopefully be answered in the following years.

References

1. Zangrossi H Jr, Viana MB, Graeff FG (1999) Anxiolytic effect of intra-amygdala injection of midazolam and 8-hydroxy-2-(di-*n*-propylamino)tetralin in the elevated T-maze. Eur J Pharmacol 369:267–270
2. VdP Soares, Zangrossi H Jr (2004) Involvement of 5-HT1A and 5-HT2 receptors of the dorsal periaqueductal gray in the regulation of the defensive behaviors generated by the elevated T-maze. Brain Res Bull 64:181–188
3. VdP Soares, Zangrossi H Jr (2009) Stimulation of 5-HT1A or 5-HT2A receptors in the ventrolateral periaqueductal gray causes anxiolytic-, but not panicolytic-like effect in rats. Behav Brain Res 197:178–185
4. Hogg S, Andrews N, File SE (1994) Contrasting behavioural effects of 8-OHDPAT in the dorsal raphe nucleus and ventral hippocampus. Neuropharmacology 33:343–348
5. Cervo L, Mocaër E, Bertaglia A, Samanin R (2000) Roles of 5-HT_{1A} receptors in the dorsal raphe and dorsal hippocampus in anxiety assessed by the behavioral effects of 8-OH-DPAT and S 15535 in a modified Geller–Seifter conflict model. Neuropharmacology 39:1037–1043
6. Viana MB, Zangrossi H Jr, Onusic GM (2008) 5-HT1A receptors of the lateral septum regulate inhibitory avoidance but not escape behavior in rats. Pharmacol Biochem Behav 89:360–366

7. Cheeta S, Kenny PJ, File SE (2000) Hippocampal and septal injections of nicotine and 8-OH-DPAT distinguish among different animal tests of anxiety. Prog Neuropsychopharmacol Biol Psychiatry 24:1053–1067
8. Menard J, Treit D (1998) The septum and the hippocampus differentially mediate anxiolytic effects of R(+)-8-OH-DPAT. Behav Pharmacol 9:93–101
9. Monassi CR, Menescal-de-Oliveira L (2004) Serotonin 5-HT_2 and 5-HT_{1A} receptors in the periaqueductal gray matter differentially modulate tonic immobility in guinea pig. Brain Res 1009:169–180
10. Nunes-de-Souza V, Nunes-de-Souza RL, Rodgers RJ, Canto-de-Souza A (2008) 5-HT_2 receptor activation in the midbrain periaqueductal gray (PAG) reduces anxiety-like behaviour in mice. Behav Brain Res 187:72–79
11. Macedo CE, Martinez RCR, Albrechet-Souza L, Molina VA, Brandão ML (2007) 5-HT_2- and D_1-mechanisms of the basolateral nucleus of the amygdala enhance conditioned fear and impair unconditioned fear. Behav Brain Res 177:100–108
12. Christianson JP, Ragole T, Amat J, Greenwood BN, Strong PV, Paul ED, Fleshner M, Watkins LR, Maier SF (2010) 5-hydroxytryptamine 2C receptors in the basolateral amygdala are involved in the expression of anxiety after uncontrollable traumatic stress. Biol Psychiatry 67:339–345
13. Alves SH, Pinheiro G, Motta V, Landeira-Fernandez J, Cruz AP (2004) Anxiogenic effects in the rat elevated plus-maze of 5-HT(2C) agonists into ventral but not dorsal hippocampus. Behav Pharmacol 15:37–43
14. Calizo LH, Ma X, Pan Y, Lemos J, Craige C, Heemstra L, Beck SG (2011) Raphe serotonin neurons are not homogenous: electrophysiological, morphological and neurochemical evidence. Neuropsychopharmacology (in press)
15. Crawford LK, Craige CP, Beck SG (2010) Increased intrinsic excitability of lateral wing serotonin neurons of the dorsal raphe: a mechanism for selective activation in stress circuits. J Neurophysiol 103:2652–2663
16. Osterberg N, Wiehle M, Oehlke O, Heidrich S, Xu C, Fan C-M, Krieglstein K, Roussa E (2011) Sim1 is a novel regulator in the differentiation of mouse dorsal raphe serotonergic neurons. Plos One 6:e19239
17. Beyer CE, Ghavami A, Lin Q, Sung A, Rhodes KJ, Dawson LA, Schechter LE, Young KH (2004) Regulators of G-protein signaling 4: Modulation of 5-HT_{1A}-mediated neurotransmitter release in vivo. Brain Res 1022:214–220
18. Liu C, Maejima T, Wyler SC, Casadesus G, Herlitze S, Deneris ES (2010) *Pet-1* is required across different stages of life to regulate serotonergic function. Nat Neurosci 13:1190–1198
19. Campbell BM, Merchant KM (2003) Serotonin 2C receptors within the basolateral amygdala induce acute fear-like responses in an open-field environment. Brain Res 993:1–9
20. Dracheva S, Lyddon R, Barley K, Marcus SM, Hurd YL, Byne WM (2009) Editing of serotonin 2C receptor mRNA in the prefrontal cortex characterizes high-novelty locomotor response behavioral trait. Neuropsychopharmacology 34:2237–2251
21. Heisler LK, Zhou L, Bajwa P, Hsu J, Tecott LH (2007) Serotonin 5-HT_{2C} receptors regulate anxiety-like behavior. Genes, Brain Behav 6:491–496
22. Kimura A, Stevenson PL, Carter RN, MacColl G, French KL, Simons JP, Al-Shawi R, Kelly V, Chapman KE, Holmes MC (2009) Overexpression of 5-HT_{2C} receptors in forebrain leads to elevated anxiety and hypoactivity. Eur J Neurosci 30:299–306
23. Gross C, Zhuang X, Stark K, Ramboz S, Oosting R, Kirby L, Santarelli L, Beck S, Hen R (2002) $Serotonin_{1A}$ receptor acts during development to establish normal anxiety-like behaviour in the adult. Nature 416:396–400
24. Gozzi A, Jain A, Giovanelli A, Bertollini C, Crestan V, Schwarz AJ, Tsetsenis T, Ragozzino D, Gross CT, Bifone A (2010) A neural switch for active and passive fear. Neuron 67:656–666
25. Tsetsenis T, Ma XH, Lo Iacono L, Beck SG, Gross C (2007) Suppression of conditioning to ambiguous cues by pharmacogenetic inhibition of the dentate gyrus. Nat Neurosci 10:896–902

Index

5

5,7-dihydroxytryptamine, 42, 80
5-HT$_{1A}$ receptors, 1, 6, 7, 22–25, 45–47, 50, 52, 56, 80, 23, 87, 90, 91, 93–95
 5-HT$_{1A}$R knockdown
 5-HT$_{1A}$R knockouts, 26, 23, 24
 8-OH-DPAT, 24, 39, 42, 46, 56, 79, 81, 82, 91, 95
 buspirone, 1
 effector coupling, 24
 NAN-190, 52
 regulation of expression by corticosteroids, 24
 regulation of expression by Freud transcription factors, 24
 role in basolateral amygdala, 42
 role in hippocampus, 45, 46
 role in prefrontal cortex, 40
 S-15535, 7
 WAY 100635, 45, 47, 56, 57, 81, 90, 105
5-HT$_{1B}$ receptors, 22, 25, 26, 87, 92, 95
 5-HT$_{1B}$R knockouts, 25
 5-HT$_{1B}$R overexpression, 25
 effector coupling
 regulation of expression by S100A10/p11, 25, 26
5-HT$_{2C}$ receptors, 22, 26, 40, 42, 45, 46, 52, 56, 59
 5-HT$_{2C}$R knockouts, 50
 agonist-directed trafficking of receptor stimulus, 26
 effector coupling, 42
 RNA editing, 26
 role in basolateral amygdala, 42
 role in hippocampus, 45
 role in prefrontal cortex, 40

A

Airjet escape, 38, 58, 59
Anxiety versus fear, 1
 affordances, 1, 2, 5
 pharmacological dissociability
 stimulus control, 3
 stressor controllability, 81, 93, 4, 6, 17, 25, 26, 37, 39, 40, 53, 54, 81, 90, 92–94
Arousal, 51, 52, 57
Attention, 57, 58
Autonomic nervous system, 37, 39, 46, 5, 7, 92, 93

B

BALB/c mice, 17, 40
Bar-press escape, 79
Bed nucleus of the stria terminalis, 93
Behavioral inhibition system, 37, 47
 activation of mPFC in anticipatory anxiety, 38
 basolateral amygdala, 37, 38, 40–42, 45, 48, 49, 79, 81, 90, 95
 functional lateralization, 38, 41, 48, 54
 Habenula, 26, 37, 46, 47, 53, 90, 92, 93, 95
 medial prefrontal cortex, 7, 18, 21, 22, 25, 26, 37–40, 44, 45, 47, 48, 58, 59, 81, 82, 87, 89, 90, 91, 93, 95
 mPFC lesions, 38
 septo-hippocampal system, 7, 20, 22, 24, 26, 37, 38, 42, 43, 45–47, 81, 91
 theta oscillations, 43, 44, 45, 91
Behavioral syndrome. *Definition of anxiety: Coping*
Benzodiazepines, 40, 49, 82

C. Maximino, *Serotonin and Anxiety*, SpringerBriefs in Neuroscience,
DOI: 10.1007/978-1-4614-4048-2, © The Author(s) 2012

C
Caffeine, 4, 41, 42, 51, 54, 59
Caudate-putamen, 90
Cell lineage
  CHO, 19, 26
  COS-7, 25
  HEK293, 18
  HeLa, 18, 19
  imortalized lymphoblast B, 21
  MN9D, 95
  RBL-2H3, 18
  RN46A, 18
Central amygdala, 18, 48, 58, 81, 89, 92
Cerebral aversive system, 1, 5, 47, 106
  bed nucleus of the stria terminalis, 4, 7, 38, 40, 50, 57, 87, 90
  central amygdala, 4, 7, 12, 37, 38, 40, 48–51, 56–58, 81, 87, 89, 90, 93
  medial hypothalamic defense system, 7, 22
  medial hypothalamus, 22, 47, 51, 81, 87
  inferior colliculus, 41, 53
  periaqueductal gray area, 6, 17, 26, 46–48, 50–58, 79, 80, 87, 93–95, 105
  modular organization, 47
  functional lateralization
  rostromedial tegmental nucleus, 7, 38, 53

C
Cholecystokinin, 39, 40, 54, 59, 81, 82, 91
Chronic unpredictable stress, 4, 16, 89
Clinical efficacy, 1, 18
Conditioned conflict paradigms, 42, 79, 81
Corticosteroids, 15, 21, 22, 24, 4, 42, 49, 52
CRF, corticotropin releasing factor, 4, 8, 16, 39, 50, 51, 52, 81, 89, 90, 92, 94, 49, 50, 52, 57, 89, 94
CRF, corticotropin-releasing factor

D
Defensive burying, 6
Definition of anxiety, 2
  action readiness, 2
  approach-avoidance conflict, 81
  coping, 4, 6–8, 38, 39, 41, 46, 53, 54, 92
  novelty, 2, 23
  risk assessment, 2, 56
  stress reactivity, 6, 26, 40
Definition of fear
  escape responses, 5, 7, 41, 46, 53, 54, 58
  fear-induced analgesia, 5, 53, 56, 57
  fear-potentiated startle, 5, 25, 49, 50
  freezing, 5–7, 21, 25, 38, 39, 41, 45, 46, 49–51, 53–58, 81, 95
  pavlovian fear conditioning, 5, 6, 25, 41, 44–46, 48–51, 54, 56, 92, 93
  tonic immobility, 53, 54, 56, 57, 94
  ultrasonic vocalization, 7, 38, 41, 58, 91
Developmental genes, 105
  gata2, 95
  gata3
  lmx1b, 16, 91
  pet1, 16, 17, 91, 94, 95
  sim1, 16, 24, 95
Dl-pCPA, 17, 79
Dorsal raphe nucleus, 8, 24, 25, 39, 42, 45, 47, 49, 50, 53, 80, 81, 82, 92, 105
  caudal portion, 16, 41, 50, 87, 91–93
  dorsal portion, 39–41, 50, 80, 87, 89, 90–93
  interfascicular portion, 39, 87
  lateral wings, 16, 52, 87, 89, 93–95
  rostral portion, 87, 92
  ventral portion, 80, 87, 89, 93, 94
Drug withdrawal

E
Elevated plus-maze, 2, 21, 22, 24, 25, 3, 38, 4, 40–42, 44–46, 48, 49, 51–54, 56, 58, 80, 81, 89, 90, 95
  "Trial 2"
Elevated T-maze, 39, 42, 46, 53, 81, 87, 89
Elevated zero-maze, 21, 95
Endocannabinoids, 52, 82, 89, 90

F
Fear potentiation of anxiety, 4
FG-7142, 38, 41, 42, 48, 51, 55, 58, 59
Forced swim, 24, 25, 54, 58, 6, 7, 92, 94

H
HAB/LAB rats, 38, 51, 58
Holeboard test, 21
Hypothalamic-pituitary-adrenal axis, 4, 5, 7, 21, 39, 46, 51, 52
Hypothalamic-pituitary-adrenal system, 39, 51, 52, 92, 95

K
Kindling, 41, 42, 48

L
Learned helplessness.
Stressor controllability, 4
Locus coeruleus, 25, 4, 57, 58, 59, 91

M
Marble burying test, 17, 24, 91
mCPP, 26, 4, 42, 50, 51, 54, 59
Monoamine oxidase, 17, 6
cellular redox state, 17, 15
isoforms, 17
MAO A knockout, 18
MAO inhibitors, 6, 18
serotonin metabolism, 17

N
Neural switch, 41, 51, 56, 7, 8
Nitric oxide, 5, 18–20, 22, 52, 53, 55, 56, 59, 82, 90, 92–94
NMDA, 50
NMDA receptor, 4, 41, 92, 94
Norepinephrine, 22, 4, 53, 57–59
Novelty-suppressed feeding, 25, 91
Novelty-suppressed feeding test, 25, 91, 45
Nucleus accumbens, 25, 26, 28, 38, 40, 7, 89, 90

O
Open-field test, 17, 2, 22, 24–26, 38, 39, 41, 42, 44, 50, 51, 81, 89, 90, 94
Opioids, 54, 56, 57
Oxytocin, 49–52

P
Post-traumatic stress disorder, 4
Predator imminence. *Consulte stimulus control*, 6
Predator stress, 39, 4, 41, 48, 51, 53, 54, 58, 81
VH-BLA plasticity, 41
Prolactin, 52

R
Renin, 52
Resident-intruder paradigm, 6, 7, 95
Restraint stress, 4, 52, 58, 81, 94
Roman RHA/LHA rats, 58
Rostroventrolateral medulla, 56, 57

S
Serotonergic "tone", 7, 20, 24, 25
Social defeat, 16, 20, 25, 37, 4, 51
Social interaction test, 17, 21, 41, 45, 46, 59, 81, 94
Sodium lactate, 52, 59, 94
Spontaneous non-ambulatory motor activity, 4, 52
Substantia nigra, 37, 53, 90

T
Tail suspension test, 24, 25
Tryptophan hydroxylase, 15, 16, 20, 48, 87, 91–93, 95
cellular redox state, 15
polymorphisms, 17, 20, 21
regulation by phosphorylation, 15
serotonin synthesis, 15
tetrahydrobiopterin and serotonin synthesis, 15

U
$Uptake_1$
polymorphisms, 21
regulation by BDNF, 21
regulation by p38 MAPK, 18–21
regulation by PP2A, 18
regulation by syntaxin-1A, 19, 20
selective serotonin reuptake inhibitors, 1, 6, 20, 40, 50, 52, 59, 81
$Uptake_2$, 18, 21, 22
organic cation transporter 3, 22, 48, 52, 87
plasma membrane monoamine transporter, 21, 22, 52, 87
Urocortin, 8, 81, 42, 89, 91, 92

V
Vasopressin, 49, 50, 52, 58
Ventral tegmental area, 37, 40, 53, 90
vGlut3, 90, 91

W
Wistar-Zagreb 5-HT rats, 20, 21

Y
Yohimbine, 41, 42, 58, 59

Printed by Books on Demand, Germany